Undergraduate Texts in Mathematics

Editors

S. Axler
F.W. Gehring
K.A. Ribet

Springer
New York
Berlin
Heidelberg
Hong Kong
London
Milan
Paris
Tokyo

Undergraduate Texts in Mathematics

Abbott: Understanding Analysis.

Anglin: Mathematics: A Concise History and Philosophy.
Readings in Mathematics.

Anglin/Lambek: The Heritage of Thales.
Readings in Mathematics.

Apostol: Introduction to Analytic Number Theory. Second edition.

Armstrong: Basic Topology.

Armstrong: Groups and Symmetry.

Axler: Linear Algebra Done Right. Second edition.

Beardon: Limits: A New Approach to Real Analysis.

Bak/Newman: Complex Analysis. Second edition.

Banchoff/Wermer: Linear Algebra Through Geometry. Second edition.

Berberian: A First Course in Real Analysis.

Bix: Conics and Cubics: A Concrete Introduction to Algebraic Curves.

Brémaud: An Introduction to Probabilistic Modeling.

Bressoud: Factorization and Primality Testing.

Bressoud: Second Year Calculus.
Readings in Mathematics.

Brickman: Mathematical Introduction to Linear Programming and Game Theory.

Browder: Mathematical Analysis: An Introduction.

Buchmann: Introduction to Cryptography.

Buskes/van Rooij: Topological Spaces: From Distance to Neighborhood.

Callahan: The Geometry of Spacetime: An Introduction to Special and General Relativity.

Carter/van Brunt: The Lebesgue–Stieltjes Integral: A Practical Introduction.

Cederberg: A Course in Modern Geometries. Second edition.

Childs: A Concrete Introduction to Higher Algebra. Second edition.

Chung: Elementary Probability Theory with Stochastic Processes. Third edition.

Cox/Little/O'Shea: Ideals, Varieties, and Algorithms. Second edition.

Croom: Basic Concepts of Algebraic Topology.

Curtis: Linear Algebra: An Introductory Approach. Fourth edition.

Devlin: The Joy of Sets: Fundamentals of Contemporary Set Theory. Second edition.

Dixmier: General Topology.

Driver: Why Math?

Ebbinghaus/Flum/Thomas: Mathematical Logic. Second edition.

Edgar: Measure, Topology, and Fractal Geometry.

Elaydi: An Introduction to Difference Equations. Second edition.

Erdős/Surányi: Topics in the Theory of Numbers.

Estep: Practical Analysis in One Variable.

Exner: An Accompaniment to Higher Mathematics.

Exner: Inside Calculus.

Fine/Rosenberger: The Fundamental Theory of Algebra.

Fischer: Intermediate Real Analysis.

Flanigan/Kazdan: Calculus Two: Linear and Nonlinear Functions. Second edition.

Fleming: Functions of Several Variables. Second edition.

Foulds: Combinatorial Optimization for Undergraduates.

Foulds: Optimization Techniques: An Introduction.

Franklin: Methods of Mathematical Economics.

Frazier: An Introduction to Wavelets Through Linear Algebra.

(continued after index)

L. Lovász
J. Pelikán
K. Vesztergombi

Discrete Mathematics

Elementary and Beyond

With 95 Illustrations

 Springer

L. Lovász
Microsoft Corporation
Microsoft Research
One Microsoft Way
Redmond, WA 98052-6399
USA
lovasz@microsoft.com

J. Pelikán
Department of Algebra
  and Number Theory
Eőtvős Loránd University
Pázmány Péter Sétany 1/C
Budapest H-1117
Hungary
pelikan@cs.elte.hu

K. Vesztergombi
Department of Mathematics
University of Washington
Box 354-350
Seattle, WA 98195-4350
USA
veszter@math.washington.edu

Mathematics Subject Classification (2000): 28-01, 30-01

Library of Congress Cataloging-in-Publication Data
Lovász, László, 1948–
    Discrete mathematics /László Lovász, József Pelikán, Katalin L. Vesztergombi.
      p. cm. — (Undergraduate texts in mathematics)
    Includes bibliographical references and index.
    ISBN 0-387-95584-4 (alk. paper) — ISBN 0-387-95585-2 (pbk. : alk. paper)
      1. Mathematics. 2. Computer science—Mathematics. I. Pelikán, József.
    II. Vesztergombi, Katalin L. III. Title. III. Series.
    QA39.3 .L68 2003
    510—dc21   2002030585

ISBN 0-387-95584-4 (hardcover)   Printed on acid-free paper.
ISBN 0-387-95585-2 (softcover)

Printed in the United States of America.

9 8 7 6 5 4 3 2 1   SPIN 10892831 (hardcover)   SPIN 10892849 (softcover)

Typesetting: Pages created by the authors using a Springer T$_{\rm E}$X macro package.

www.springer-ny.com

Springer-Verlag New York Berlin Heidelberg
A member of BertelsmannSpringer Science+Business Media GmbH

Preface

For most students, the first and often only course in college mathematics is calculus. It is true that calculus is the single most important field of mathematics, whose emergence in the seventeenth century signaled the birth of modern mathematics and was the key to the successful applications of mathematics in the sciences and engineering.

But calculus (or analysis) is also very technical. It takes a lot of work even to introduce its fundamental notions like continuity and the derivative (after all, it took two centuries just to develop the proper definition of these notions). To get a feeling for the power of its methods, say by describing one of its important applications in detail, takes years of study.

If you want to become a mathematician, computer scientist, or engineer, this investment is necessary. But if your goal is to develop a feeling for what mathematics is all about, where mathematical methods can be helpful, and what kinds of questions do mathematicians work on, you may want to look for the answer in some other fields of mathematics.

There are many success stories of applied mathematics outside calculus. A recent hot topic is mathematical cryptography, which is based on number theory (the study of the positive integers $1, 2, 3, \ldots$), and is widely applied, for example, in computer security and electronic banking. Other important areas in applied mathematics are linear programming, coding theory, and the theory of computing. The mathematical content in these applications is collectively called *discrete mathematics*. (The word "discrete" is used in the sense of "separated from each other," the opposite of "continuous;" it is also often used in the more restrictive sense of "finite." The more everyday version of this word, meaning "circumspect," is spelled "discreet.")

The aim of this book is not to cover "discrete mathematics" in depth (it should be clear from the description above that such a task would be ill-defined and impossible anyway). Rather, we discuss a number of selected results and methods, mostly from the areas of combinatorics and graph theory, with a little elementary number theory, probability, and combinatorial geometry.

It is important to realize that there is no mathematics without *proofs*. Merely stating the facts, without saying something about why these facts are valid, would be terribly far from the spirit of mathematics and would make it impossible to give any idea about how it works. Thus, wherever possible, we will give the proofs of the theorems we state. Sometimes this is not possible; quite simple, elementary facts can be extremely difficult to prove, and some such proofs may take advanced courses to go through. In these cases, we will at least state that the proof is highly technical and goes beyond the scope of this book.

Another important ingredient of mathematics is *problem solving*. You won't be able to learn any mathematics without dirtying your hands and trying out the ideas you learn about in the solution of problems. To some, this may sound frightening, but in fact, most people pursue this type of activity almost every day: Everybody who plays a game of chess or solves a puzzle is solving discrete mathematical problems. The reader is strongly advised to answer the questions posed in the text and to go through the problems at the end of each chapter of this book. Treat it as puzzle solving, and if you find that some idea that you came up with in the solution plays some role later, be satisfied that you are beginning to get the essence of how mathematics develops.

We hope that we can illustrate that mathematics is a building, where results are built on earlier results, often going back to the great Greek mathematicians; that mathematics is alive, with more new ideas and more pressing unsolved problems than ever; and that mathematics is also an art, where the beauty of ideas and methods is as important as their difficulty or applicability.

László Lovász   József Pelikán   Katalin Vesztergombi

Contents

1
Let's Count!

1.1 A Party

Alice invites six guests to her birthday party: Bob, Carl, Diane, Eve, Frank, and George. When they arrive, they shake hands with each other (strange European custom). This group is strange anyway, because one of them asks, "How many handshakes does this mean?"

"I shook 6 hands altogether," says Bob, "and I guess, so did everybody else."

"Since there are seven of us, this should mean $7 \cdot 6 = 42$ handshakes," ventures Carl.

"This seems too many" says Diane. "The same logic gives 2 handshakes if two persons meet, which is clearly wrong."

"This is exactly the point: Every handshake was counted twice. We have to divide 42 by 2 to get the right number: 21," with which Eve settles the issue.

When they go to the table, they have a difference of opinion about who should sit where. To resolve this issue, Alice suggests, "Let's change the seating every half hour, until we get every seating."

"But you stay at the head of the table," says George, "since it is your birthday."

How long is this party going to last? How many different seatings are there (with Alice's place fixed)?

Let us fill the seats one by one, starting with the chair on Alice's right. Here we can put any of the 6 guests. Now look at the second chair. If Bob sits in the first chair, we can put any of the remaining 5 guests in the second chair; if Carl sits in the first chair, we again have 5 choices for the second chair, etc. Each of the six choices for the first chair gives us five choices for the second chair, so the number of ways to fill the first two chairs is $5 + 5 + 5 + 5 + 5 + 5 = 6 \cdot 5 = 30$. Similarly, no matter how we fill the first two chairs, we have 4 choices for the third chair, which gives $6 \cdot 5 \cdot 4$ ways to fill the first three chairs. Proceeding similarly, we find that the number of ways to seat the guests is $6 \cdot 5 \cdot 4 \cdot 3 \cdot 2 \cdot 1 = 720$.

If they change seats every half hour, it will take 360 hours, that is, 15 days, to go through all the seating arrangements. Quite a party, at least as far as the duration goes!

1.1.1 How many ways can these people be seated at the table if Alice, too, can sit anywhere?

After the cake, the crowd wants to dance (boys with girls, remember, this is a conservative European party). How many possible pairs can be formed?

OK, this is easy: there are 3 girls, and each can choose one of 4 boys, this makes $3 \cdot 4 = 12$ possible pairs.

After ten days have passed, our friends really need some new ideas to keep the party going. Frank has one: "Let's pool our resources and win the lottery! All we have to do is to buy enough tickets so that no matter what they draw, we will have a ticket with the winning numbers. How many tickets do we need for this?"

(In the lottery they are talking about, 5 numbers are selected out of 90.)

"This is like the seating," says George. "Suppose we fill out the tickets so that Alice marks a number, then she passes the ticket to Bob, who marks a number and passes it to Carl, and so on. Alice has 90 choices, and no matter what she chooses, Bob has 89 choices, so there are $90 \cdot 89$ choices for the first two numbers, and going on similarly, we get $90 \cdot 89 \cdot 88 \cdot 87 \cdot 86$ possible choices for the five numbers."

"Actually, I think this is more like the handshake question," says Alice. "If we fill out the tickets the way you suggested, we get the same ticket more then once. For example, there will be a ticket where I mark 7 and Bob marks 23, and another one where I mark 23 and Bob marks 7."

Carl jumps up: "Well, let's imagine a ticket, say, with numbers $7, 23, 31, 34,$ and 55. How many ways do we get it? Alice could have marked any of them; no matter which one it was that she marked, Bob could have marked any of the remaining four. Now this is really like the seating problem. We get every ticket $5 \cdot 4 \cdot 3 \cdot 2 \cdot 1$ times."

"So," concludes Diane, "if we fill out the tickets the way George proposed, then among the $90 \cdot 89 \cdot 88 \cdot 87 \cdot 86$ tickets we get, every 5-tuple occurs not only once, but $5 \cdot 4 \cdot 3 \cdot 2 \cdot 1$ times. So the number of *different* tickets is only

$$\frac{90 \cdot 89 \cdot 88 \cdot 87 \cdot 86}{5 \cdot 4 \cdot 3 \cdot 2 \cdot 1}.$$

We only need to buy this number of tickets."

Somebody with a good pocket calculator computed this value in a twinkling; it was 43,949,268. So they had to decide (remember, this happens in a poor European country) that they didn't have enough money to buy so many tickets. (Besides, they would win much less. And to fill out so many tickets would spoil the party!)

So they decide to play cards instead. Alice, Bob, Carl and Diane play bridge. Looking at his cards, Carl says, "I think I had the same hand last time."

"That is very unlikely" says Diane.

How unlikely is it? In other words, how many different hands can you have in bridge? (The deck has 52 cards, each player gets 13.) We hope you have noticed that this is essentially the same question as the lottery problem. Imagine that Carl picks up his cards one by one. The first card can be any one of the 52 cards; whatever he picked up first, there are 51 possibilities for the second card, so there are $52 \cdot 51$ possibilities for the first two cards. Arguing similarly, we see that there are $52 \cdot 51 \cdot 50 \cdots 40$ possibilities for the 13 cards.

But now every hand has been counted many times. In fact, if Eve comes to kibitz and looks into Carl's cards after he has arranged them and tries to guess (we don't know why) the order in which he picked them up, she could think, "He could have picked up any of the 13 cards first; he could have picked up any of the remaining 12 cards second; any of the remaining 11 cards third.... Aha, this is again like the seating: There are $13 \cdot 12 \cdots 2 \cdot 1$ orders in which he could have picked up his cards."

But this means that the number of *different* hands in bridge is

$$\frac{52 \cdot 51 \cdot 50 \cdots 40}{13 \cdot 12 \cdots 2 \cdot 1} = 635{,}013{,}559{,}600.$$

So the chance that Carl had the same hand twice in a row is one in 635,013,559,600, which is very small indeed.

Finally, the six guests decide to play chess. Alice, who just wants to watch, sets up three boards.

"How many ways can you guys be matched with each other?" she wonders. "This is clearly the same problem as seating you on six chairs; it does not matter whether the chairs are around the dinner table or at the three boards. So the answer is 720 as before."

"I think you should not count it as a different pairing if two people at the same board switch places," says Bob, "and it shouldn't matter which pair sits at which board."

"Yes, I think we have to agree on what the question really means," adds Carl. "If we include in it who plays white on each board, then if a pair switches places we do get a different matching. But Bob is right that it doesn't matter which pair uses which board."

"What do you mean it does not matter? You sit at the first board, which is closest to the peanuts, and I sit at the last, which is farthest," says Diane.

"Let's just stick to Bob's version of the question" suggests Eve. "It is not hard, actually. It is like with handshakes: Alice's figure of 720 counts every pairing several times. We could rearrange the 3 boards in 6 different ways, without changing the pairing."

"And each pair may or may not switch sides" adds Frank. "This means $2 \cdot 2 \cdot 2 = 8$ ways to rearrange people without changing the pairing. So in fact, there are $6 \cdot 8 = 48$ ways to sit that all mean the same pairing. The 720 seatings come in groups of 48, and so the number of matchings is $720/48 = 15$."

"I think there is another way to get this," says Alice after a little time. "Bob is youngest, so let him choose a partner first. He can choose his partner in 5 ways. Whoever is youngest among the rest can choose his or her partner in 3 ways, and this settles the pairing. So the number of pairings is $5 \cdot 3 = 15$."

"Well, it is nice to see that we arrived at the same figure by two really different arguments. At the least, it is reassuring" says Bob, and on this happy note we leave the party.

1.1.2 What is the number of pairings in Carl's sense (when it matters who sits on which side of the board, but the boards are all alike), and in Diane's sense (when it is the other way around)?

1.1.3 What is the number of pairings (in all the various senses as above) in a party of 10?

1.2 Sets and the Like

We want to formalize assertions like "the problem of counting the number of hands in bridge is essentially the same as the problem of counting tickets in the lottery." The most basic tool in mathematics that helps here is the notion of a *set*. Any collection of distinct objects, called *elements*, is a set. The deck of cards is a set, whose elements are the cards. The participants in the party form a set, whose elements are Alice, Bob, Carl, Diane, Eve, Frank, and George (let us denote this set by P). Every lottery ticket of the type mentioned above contains a set of 5 numbers.

For mathematics, various sets of numbers are especially important: the set of real numbers, denoted by $\mathbb{R}$; the set of rational numbers, denoted by $\mathbb{Q}$; the set of integers, denote by $\mathbb{Z}$; the set of non-negative integers, denoted by $\mathbb{Z}_+$; the set of positive integers, denoted by $\mathbb{N}$. The *empty set*, the set with no elements, is another important (although not very interesting) set; it is denoted by $\emptyset$.

If A is a set and b is an element of A, we write $b \in A$. The number of elements of a set A (also called the *cardinality* of A) is denoted by $|A|$. Thus $|P| = 7$, $|\emptyset| = 0$, and $|\mathbb{Z}| = \infty$ (infinity).[1]

We may specify a set by listing its elements between braces; so

$$P = \{\text{Alice, Bob, Carl, Diane, Eve, Frank, George}\}$$

is the set of participants in Alice's birthday party, and

$$\{12, 23, 27, 33, 67\}$$

is the set of numbers on my uncle's lottery ticket. Sometimes, we replace the list by a verbal description, like

$$\{\text{Alice and her guests}\}.$$

Often, we specify a set by a property that singles out the elements from a large "universe" like that of all real numbers. We then write this property inside the braces, but after a colon. Thus

$$\{x \in \mathbb{Z} : \ x \geq 0\}$$

is the set of non-negative integers (which we have called $\mathbb{Z}_+$ before), and

$$\{x \in P : \ x \text{ is a girl}\} = \{\text{Alice, Diane, Eve}\}$$

(we will denote this set by G). Let us also tell you that

$$\{x \in P : \ x \text{ is over 21 years old}\} = \{\text{Alice, Carl, Frank}\}$$

(we will denote this set by D).

A set A is called a *subset* of a set B if every element of A is also an element of B. In other words, A consists of certain elements of B. We can allow A to consist of all elements of B (in which case $A = B$) or none of them (in which case $A = \emptyset$), and still consider it a subset. So the empty set is a subset of every set. The relation that A is a subset of B is denoted by $A \subseteq B$. For example, among the various sets of people considered above, $G \subseteq P$ and $D \subseteq P$. Among the sets of numbers, we have a long chain:

$$\emptyset \subseteq \mathbb{N} \subseteq \mathbb{Z}_+ \subseteq \mathbb{Z} \subseteq \mathbb{Q} \subseteq \mathbb{R}.$$

---

[1] In mathematics one can distinguish various levels of "infinity"; for example, one can distinguish between the cardinalities of $\mathbb{Z}$ and $\mathbb{R}$. This is the subject matter of *set theory* and does not concern us here.

The notation $A \subset B$ means that A is a subset of B but not all of B. In the chain above, we could replace the $\subseteq$ signs by $\subset$.

If we have two sets, we can define various other sets with their help. The *intersection* of two sets is the set consisting of those elements that are elements of both sets. The intersection of two sets A and B is denoted by $A \cap B$. For example, we have $G \cap D = \{\text{Alice}\}$. Two sets whose intersection is the empty set (in other words, they have no element in common) are called *disjoint*.

The *union* of two sets is the set consisting of those elements that are elements of at least one of the sets. The union of two sets A and B is denoted by $A \cup B$. For example, we have $G \cup D = \{\text{Alice, Carl, Diane, Eve, Frank}\}$.

The *difference* of two sets A and B is the set of elements that belong to A but not to B. The difference of two sets A and B is denoted by $A \setminus B$. For example, we have $G \setminus D = \{\text{Diane, Eve}\}$.

The *symmetric difference* of two sets A and B is the set of elements that belong to exactly one of A and B. The symmetric difference of two sets A and B is denoted by $A \triangle B$. For example, we have $G \triangle D = \{\text{Carl, Diane, Eve, Frank}\}$.

Intersection, union, and the two kinds of differences are similar to addition, multiplication, and subtraction. However, they are operations on *sets*, rather than operations on *numbers*. Just like operations on numbers, set operations obey many useful rules (identities). For example, for any three sets A, B, and C,

$$A \cap (B \cup C) = (A \cap B) \cup (A \cap C). \tag{1.1}$$

To see that this is so, think of an element x that belongs to the set on the left-hand side. Then we have $x \in A$ and also $x \in B \cup C$. This latter assertion is the same as saying that either $x \in B$ or $x \in C$. If $x \in B$, then (since we also have $x \in C$) we have $x \in A \cap B$. If $x \in C$, then similarly we get $x \in A \cap C$. So we know that $x \in A \cap B$ or $x \in A \cap C$. By the definition of the union of two sets, this is the same as saying that $x \in (A \cap B) \cup (A \cap C)$.

Conversely, consider an element that belongs to the right-hand side. By the definition of union, this means that $x \in A \cap B$ or $x \in A \cap C$. In the first case, we have $x \in A$ and also $x \in B$. In the second, we get $x \in A$ and also $x \in C$. So in either case $x \in A$, and we either have $x \in B$ or $x \in C$, which implies that $x \in B \cup C$. But this means that $A \cap (B \cup C)$.

This kind of argument gets a bit boring, even though there is really nothing to it other than simple logic. One trouble with it is that it is so lengthy that it is easy to make an error in it. There is a nice graphic way to support such arguments. We represent the sets A, B, and C by three overlapping circles (Figure 1.1). We imagine that the common elements of A, B, and C are put in the common part of the three circles; those elements of A that are also in B but not in C are put in the common part of circles A and B outside C, etc. This drawing is called the *Venn diagram* of the three sets.

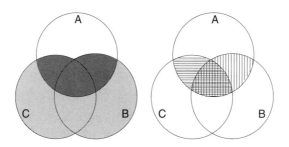

FIGURE 1.1. The Venn diagram of three sets, and the sets on both sides of (1.1).

Now, where are those elements in the Venn diagram that belong to the left-hand side of (1.1)? We have to form the union of B and C, which is the gray set in Figure 1.1(a), and then intersect it with A, to get the dark gray part. To get the set on the right-hand side, we have to form the sets $A \cap B$ and $A \cap C$ (marked by vertical and horizontal lines, respectively in Figure 1.1(b)), and then form their union. It is clear from the picture that we get the same set. This illustrates that Venn diagrams provide a safe and easy way to prove such identities involving set operations.

The identity (1.1) is nice and quite easy to remember: If we think of "union" as a sort of addition (this is quite natural), and "intersection" as a sort of multiplication (hmm... not so clear why; perhaps after we learn about probability in Chapter 5 you'll see it), then we see that (1.1) is completely analogous to the familiar distributive rule for numbers:

$$a(b + c) = ab + ac.$$

Does this analogy go any further? Let's think of other properties of addition and multiplication. Two important properties are that they are *commutative*,

$$a + b = b + a, \qquad ab = ba,$$

and *associative*,

$$(a + b) + c = a + (b + c), \qquad (ab)c = a(bc).$$

It turns out that these are also properties of the union and intersection operations:

$$A \cup B = B \cup A, \qquad A \cap B = B \cap A, \tag{1.2}$$

and

$$(A \cup B) \cup C = A \cup (B \cup C), \qquad (A \cap B) \cap C = A \cap (B \cap C). \tag{1.3}$$

The proof of these identities is left to the reader as an exercise.

Warning! Before going too far with this analogy, let us point out that there is another distributive law for sets:

$$A \cup (B \cap C) = (A \cup B) \cap (A \cup C). \tag{1.4}$$

We get this simply by interchanging "union" and "intersection" in (1.1). (This identity can be proved just like (1.1); see Exercise 1.2.16.) This second distributivity is something that has no analogue for numbers: In general,

$$a + bc \neq (a + b)(a + c)$$

for three numbers a, b, c.

There are other remarkable identities involving union, intersection, and also the two kinds of differences. These are useful, but not very deep: They reflect simple logic. So we don't list them here, but state several of these below in the exercises.

1.2.1 Name sets whose elements are (a) buildings, (b) people, (c) students, (d) trees, (e) numbers, (f) points.

1.2.2 What are the elements of the following sets: (a) army, (b) mankind, (c) library, (d) the animal kingdom?

1.2.3 Name sets having cardinality (a) 52, (b) 13, (c) 32, (d) 100, (e) 90, (f) 2,000,000.

1.2.4 What are the elements of the following (admittedly peculiar) set: $\{\text{Alice}, \{1\}\}$?

1.2.5 Is an "element of a set" a special case of a "subset of a set"?

1.2.6 List all subsets of $\{0, 1, 3\}$. How many do you get?

1.2.7 Define at least three sets of which $\{\text{Alice, Diane, Eve}\}$ is a subset.

1.2.8 List all subsets of $\{a, b, c, d, e\}$, containing a but not containing b.

1.2.9 Define a set of which both $\{1, 3, 4\}$ and $\{0, 3, 5\}$ are subsets. Find such a set with the smallest possible number of elements.

1.2.10 (a) Which set would you call the union of $\{a, b, c\}$, $\{a, b, d\}$ and $\{b, c, d, e\}$?

  (b) Find the union of the first two sets, and then the union of this with the third. Also, find the union of the last two sets, and then the union of this with the first set. Try to formulate what you have observed.

  (c) Give a definition of the union of more than two sets.

1.2.11 Explain the connection between the notion of the union of sets and Exercise 1.2.9.

1.2.12 We form the union of a set with 5 elements and a set with 9 elements. Which of the following numbers can we get as the cardinality of the union: 4, 6, 9, 10, 14, 20?

1.2.13 We form the union of two sets. We know that one of them has n elements and the other has m elements. What can we infer about the cardinality of their union?

1.2.14 What is the intersection of

(a) the sets $\{0, 1, 3\}$ and $\{1, 2, 3\}$;

(b) the set of girls in this class and the set of boys in this class;

(c) the set of prime numbers and the set of even numbers?

1.2.15 We form the intersection of two sets. We know that one of them has n elements and the other has m elements. What can we infer about the cardinality of their intersection?

1.2.16 Prove (1.2), (1.3), and (1.4).

1.2.17 Prove that $|A \cup B| + |A \cap B| = |A| + |B|$.

1.2.18   (a) What is the symmetric difference of the set $\mathbb{Z}_+$ of nonnegative integers and the set E of even integers ($E = \{\dots, -4, -2, 0, 2, 4, \dots\}$ contains both negative and positive even integers).

(b) Form the symmetric difference of A and B to get a set C. Form the symmetric difference of A and C. What did you get? Give a proof of the answer.

1.3   The Number of Subsets

Now that we have introduced the notion of subsets, we can formulate our first general combinatorial problem: What is the number of all subsets of a set with n elements?

We start with trying out small numbers. It makes no difference what the elements of the set are; we call them a, b, c etc. The empty set has only one subset (namely, itself). A set with a single element, say $\{a\}$, has two subsets: the set $\{a\}$ itself and the empty set $\emptyset$. A set with two elements, say $\{a, b\}$, has four subsets: $\emptyset, \{a\}, \{b\}$, and $\{a, b\}$. It takes a little more effort to list all the subsets of a set $\{a, b, c\}$ with 3 elements:

$$\emptyset, \{a\}, \{b\}, \{c\}, \{a, b\}, \{b, c\}, \{a, c\}, \{a, b, c\}. \tag{1.5}$$

We can make a little table from these data:

Number of elements	0	1	2	3
Number of subsets	1	2	4	8

Looking at these values, we observe that the number of subsets is a power of 2: If the set has n elements, the result is 2^n, at least on these small examples.

It is not difficult to see that this is always the answer. Suppose you have to select a subset of a set A with n elements; let us call these elements $a_1, a_2, \ldots, a_n$. Then we may or may not want to include a_1, in other words, we can make two possible decisions at this point. No matter how we decided about a_1, we may or may not want to include a_2 in the subset; this means two possible decisions, and so the number of ways we can decide about a_1 and a_2 is $2 \cdot 2 = 4$. Now no matter how we decide about a_1 and a_2, we have to decide about a_3, and we can again decide in two ways. Each of these ways can be combined with each of the 4 decisions we could have made about a_1 and a_2, which makes $4 \cdot 2 = 8$ possibilities to decide about a_1, a_2 and a_3.

We can go on similarly: No matter how we decide about the first k elements, we have two possible decisions about the next, and so the number of possibilities doubles whenever we take a new element. For deciding about all the n elements of the set, we have 2^n possibilities.

Thus we have derived the following theorem.

Theorem 1.3.1 *A set with n elements has 2^n subsets.*

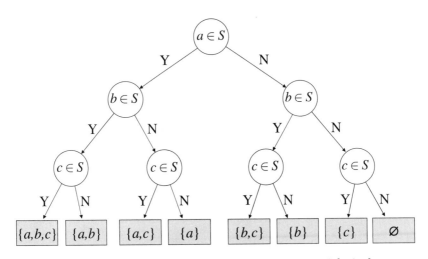

FIGURE 1.2. A decision tree for selecting a subset of $\{a, b, c\}$.

We can illustrate the argument in the proof by the picture in Figure 1.2. We read this figure as follows. We want to select a subset called S. We start from the circle on the top (called a *node*). The node contains a question: Is a an element of S? The two arrows going out of this node are labeled with the two possible answers to this question (Yes and No). We make a decision and follow the appropriate arrow (also called an *edge*) to the node at the other end. This node contains the next question: Is b an element of S? Follow the arrow corresponding to your answer to the next node, which contains the third (and in this case last) question you have to answer to determine the subset: Is c an element of S? Giving an answer and following the appropriate arrow we finally get to a node that does not represent a question, but contains a listing of the elements of S.

Thus to select a subset corresponds to walking down this diagram from the top to the bottom. There are just as many subsets of our set as there are nodes on the last level. Since the number of nodes doubles from level to level as we go down, the last level contains $2^3 = 8$ nodes (and if we had an n-element set, it would contain 2^n nodes).

Remark. A picture like this is called a *tree*. (This is not a formal definition; that will follow later.) If you want to know why the tree is growing upside down, ask the computer scientists who introduced this convention. (The conventional wisdom is that they never went out of the room, and so they never saw a real tree.)

We can give another proof of Theorem 1.3.1. Again, the answer will be made clear by asking a question about subsets. But now we don't want to select a subset; what we want is to *enumerate* subsets, which means that we want to label them with numbers $0, 1, 2, \ldots$ so that we can speak, say, about subset number 23 of the set. In other words, we want to arrange the subsets of the set in a list and then speak about the 23rd subset on the list.

(We actually want to call the first subset of the list number 0, the second subset on the list number 1 etc. This is a little strange, but this time it is the logicians who are to blame. In fact, you will find this quite natural and handy after a while.)

There are many ways to order the subsets of a set to form a list. A fairly natural thing to do is to start with $\emptyset$, then list all subsets with 1 element, then list all subsets with 2 elements, etc. This is the way the list (1.5) is put together.

Another possibility is to order the subsets as in a phone book. This method will be more transparent if we write the subsets without braces and commas. For the subsets of $\{a, b, c\}$, we get the list

$$\emptyset, a, ab, abc, ac, b, bc, c.$$

These are indeed useful and natural ways of listing all subsets. They have one shortcoming, though. Imagine the list of the subsets of 10 elements, and ask yourself to name the 233rd subset on the list, without actually writing down the whole list. This would be difficult! Is there a way to make it easier?

Let us start with another way of denoting subsets (another *encoding* in the mathematical jargon). We illustrate it on the subsets of $\{a, b, c\}$. We look at the elements one by one, and write down a 1 if the element occurs in the subset and a 0 if it does not. Thus for the subset $\{a, c\}$, we write down 101, since a is in the subset, b is not, and c is in it again. This way every subset is "encoded" by a string of length 3, consisting of 0's and 1's. If we specify any such string, we can easily read off the subset it corresponds to. For example, the string 010 corresponds to the subset $\{b\}$, since the first 0 tells us that a is not in the subset, the 1 that follows tells us that b is in there, and the last 0 tells us that c is not there.

Now, such strings consisting of 0's and 1's remind us of the *binary representation* of integers (in other words, representations in base 2). Let us recall the binary form of nonnegative integers up to 10:

$$0 = 0_2$$
$$1 = 1_2$$
$$2 = 10_2$$
$$3 = 2 + 1 = 11_2$$
$$4 = 100_2$$
$$5 = 4 + 1 = 101_2$$
$$6 = 4 + 2 = 110_2$$
$$7 = 4 + 2 + 1 = 111_2$$
$$8 = 1000_2$$
$$9 = 8 + 1 = 1001_2$$
$$10 = 8 + 2 = 1010_2$$

(We put the subscript 2 there to remind ourselves that we are working in base 2, not 10.)

Now, the binary forms of integers $0, 1, \ldots, 7$ look almost like the "codes" of subsets; the difference is that the binary form of an integer (except for 0) always starts with a 1, and the first 4 of these integers have binary forms shorter than 3, while all codes of subsets of a 3-element set consist of exactly 3 digits. We can make this difference disappear if we append 0's to the binary forms at their beginning, to make them all have the same length. This way we get the following correspondence:

$$
\begin{array}{ccccccc}
0 & \Leftrightarrow & 0_2 & \Leftrightarrow & 000 & \Leftrightarrow & \emptyset \\
1 & \Leftrightarrow & 1_2 & \Leftrightarrow & 001 & \Leftrightarrow & \{c\} \\
2 & \Leftrightarrow & 10_2 & \Leftrightarrow & 010 & \Leftrightarrow & \{b\} \\
3 & \Leftrightarrow & 11_2 & \Leftrightarrow & 011 & \Leftrightarrow & \{b, c\} \\
4 & \Leftrightarrow & 100_2 & \Leftrightarrow & 100 & \Leftrightarrow & \{a\} \\
5 & \Leftrightarrow & 101_2 & \Leftrightarrow & 101 & \Leftrightarrow & \{a, c\} \\
6 & \Leftrightarrow & 110_2 & \Leftrightarrow & 110 & \Leftrightarrow & \{a, b\} \\
7 & \Leftrightarrow & 111_2 & \Leftrightarrow & 111 & \Leftrightarrow & \{a, b, c\}
\end{array}
$$

So we see that the subsets of $\{a, b, c\}$ correspond to the numbers $0, 1, \ldots, 7$.

What happens if we consider, more generally, subsets of a set with n elements? We can argue just as we did above, to get that the subsets of an n-element set correspond to integers, starting with 0 and ending with the largest integer that has n digits in its binary representation (digits in the binary representation are usually called *bits*). Now, the smallest number with $n+1$ bits is 2^n, so the subsets correspond to numbers $0, 1, 2, \ldots, 2^n - 1$. It is clear that the number of these numbers in 2^n, and hence the number of subsets is 2^n.

Now we can answer our question about the 233rd subset of a 10-element set. We have to convert 233 to binary notation. Since 233 is odd, its last binary digit (bit) will be 1. Let us cut off this last bit. This is the same as subtracting 1 from 233 and then dividing it by 2: We get $(233 - 1)/2 = 116$. This number is even, so its last bit will be 0. Let us cut this off again; we get $(116 - 0)/2 = 58$. Again, the last bit is 0, and cutting it off we get $(58 - 0)/2 = 29$. This is odd, so its last bit is 1, and cutting it off we get $(29 - 1)/2 = 14$. Cutting off a 0, we get $(14 - 0)/2 = 7$; cutting off a 1, we get $(7 - 1)/2 = 3$; cutting off a 1, we get $(3 - 1)/2 = 1$; cutting off a 1, we get 0. So the binary form of 233 is 11101001, which corresponds to the code 0011101001.

It follows that if $a_1, \ldots, a_{10}$ are the elements of our set, then the 233rd subset of a 10-element set consists of the elements $\{a_3, a_4, a_5, a_7, a_{10}\}$.

Comments. We have given two proofs of Theorem 1.3.1. You may wonder why we needed two proofs. Certainly not because a single proof would not have given enough confidence in the truth of the statement! Unlike in a legal procedure, a mathematical proof either gives absolute certainty or else it is useless. No matter how many incomplete proofs we give, they don't add up to a single complete proof.

For that matter, we could ask you to take our word for it, and not give any proof. Later, in some cases this will be necessary, when we will state theorems whose proofs are too long or too involved to be included in this introductory book.

So why did we bother to give any proof, let alone two proofs of the same statement? The answer is that every proof reveals much more than just the bare fact stated in the theorem, and this revelation may be more valuable than the theorem itself. For example, the first proof given above introduced the idea of breaking down the selection of a subset into independent decisions and the representation of this idea by a "decision tree"; we will use this idea repeatedly.

The second proof introduced the idea of enumerating these subsets (labeling them with integers $0, 1, 2, \ldots$). We also saw an important method of counting: We established a correspondence between the objects we wanted to count (the subsets) and some other kinds of objects that we can count easily (the numbers $0, 1, \ldots, 2^n - 1$). In this correspondence:

— for every subset, we had exactly one corresponding number, and

— for every number, we had exactly one corresponding subset.

A correspondence with these properties is called a *one-to-one correspondence* (or *bijection*). If we can make a one-to-one correspondence between the elements of two sets, then they have the same number of elements.

1.3.1 Under the correspondence between numbers and subsets described above, which numbers correspond to (a) subsets with 1 element, (b) the whole set? (c) Which sets correspond to even numbers?

1.3.2 What is the number of subsets of a set with n elements, containing a given element?

1.3.3 Show that a nonempty set has the same number of odd subsets (i.e., subsets with an odd number of elements) as even subsets.

1.3.4 What is the number of integers with (a) at most n (decimal) digits; (b) exactly n digits (don't forget that there are positive and negative numbers!)?

1.4   The Approximate Number of Subsets

So, we know that the number of subsets of a 100-element set is 2^{100}. This is a large number, but how large? It would be good to know, at least, how many digits it will have in the usual decimal form. Using computers, it would not be too hard to find the decimal form of this number ($2^{100} = 1267650600228229401496703205376$), but suppose we have no computers at hand. Can we at least estimate the order of magnitude of it?

We know that $2^3 = 8 < 10$, and hence (raising both sides of this inequality to the 33rd power) $2^{99} < 10^{33}$. Therefore, $2^{100} < 2 \cdot 10^{33}$. Now $2 \cdot 10^{33}$ is a 2 followed by 33 zeros; it has 34 digits, and therefore 2^{100} has at most 34 digits.

We also know that $2^{10} = 1024 > 1000 = 10^3$; these two numbers are quite close to each other[2]. Hence $2^{100} > 10^{30}$, which means that 2^{100} has at least 31 digits.

This gives us a reasonably good idea of the size of 2^{100}. With a little more high-school math, we can get the number of digits exactly. What does it mean that a number has exactly k digits? It means that it is between 10^{k-1} and 10^k (the lower bound is allowed, the upper is not). We want to find the value of k for which

$$10^{k-1} \le 2^{100} < 10^k.$$

---

[2]The fact that 2^{10} is so close to 10^3 is used—or rather misused—in the name "kilobyte," which means 1024 bytes, although it should mean 1000 bytes, just as a "kilogram" means 1000 grams. Similarly, "megabyte" means 2^{20} bytes, which is close to 1 million bytes, but not exactly equal.

Now we can write 2^{100} in the form 10^x, only x will not be an integer: the appropriate value of x is $x = \lg 2^{100} = 100 \lg 2$ (we use lg to denote logarithm with base 10). We have then

$$k - 1 \leq x < k,$$

which means that $k - 1$ is the largest integer not exceeding x. Mathematicians have a name for this: It is the *integer part*, or *floor*, of x, and it is denoted by $\lfloor x \rfloor$. We can also say that we obtain k by rounding x down to the next integer. There is also a name for the number obtained by rounding x up to the next integer: It is called the *ceiling* of x, and denoted by $\lceil x \rceil$.

Using any scientific calculator (or table of logarithms), we see that $\lg 2 \approx 0.30103$, thus $100 \lg 2 \approx 30.103$, and rounding this down we get that $k - 1 = 30$. Thus 2^{100} has 31 digits.

1.4.1 How many bits (binary digits) does 2^{100} have if written in base 2?

1.4.2 Find a formula for the number of digits of 2^n.

1.5   Sequences

Motivated by the "encoding" of subsets as strings of 0's and 1's, we may want to determine the number of strings of length n composed of some other set of symbols, for example, a, b and c. The argument we gave for the case of 0's and 1's can be carried over to this case without any essential change. We can observe that for the first element of the string, we can choose any of a, b and c, that is, we have 3 choices. No matter what we choose, there are 3 choices for the second element of the string, so the number of ways to choose the first two elements is $3^2 = 9$. Proceeding in a similar manner, we get that the number of ways to choose the whole string is 3^n.

In fact, the number 3 has no special role here; the same argument proves the following theorem:

Theorem 1.5.1 *The number of strings of length n composed of k given elements is k^n.*

The following problem leads to a generalization of this question. Suppose that a database has 4 fields: the first, containing an 8-character abbreviation of an employee's name; the second, M or F for sex; the third, the birthday of the employee, in the format mm-dd-yy (disregarding the problem of not being able to distinguish employees born in 1880 from employees born in 1980); and the fourth, a job code that can be one of 13 possibilities. How many different records are possible?

The number will certainly be large. We already know from theorem 1.5.1 that the first field may contain $26^8 > 200,000,000,000$ names (most of these will be very difficult to pronounce, and are not likely to occur, but let's count all of them as possibilities). The second field has 2 possible entries. The third field can be thought of as three separate fields, having 12, 31, and 100 possible entries, respectively (some combinations of these will never occur, for example, 04-31-76 or 02-29-13, but let's ignore this). The last field has 13 possible entries.

Now how do we determine the number of ways these can be combined? The argument we described above can be repeated, just "3 choices" has to be replaced, in order, by "26^8 choices," "2 choices," "12 choices," "31 choices," "100 choices," and "13 choices." We get that the answer is $26^8 \cdot 2 \cdot 12 \cdot 31 \cdot 100 \cdot 13 = 201{,}977{,}536{,}857{,}907{,}200$.

We can formulate the following generalization of Theorem 1.5.1 (the proof consists of repeating the argument above).

Theorem 1.5.2 *Suppose that we want to form strings of length n by using any of a given set of k_1 symbols as the first element of the string, any of a given set of k_2 symbols as the second element of the string, etc., any of a given set of k_n symbols as the last element of the string. Then the total number of strings we can form is $k_1 \cdot k_2 \cdots k_n$.*

As another special case, consider the following problem: how many non-negative integers have exactly n digits (in decimal)? It is clear that the first digit can be any of 9 numbers $(1, 2, \ldots, 9)$, while the second, third, etc., digits can be any of the 10 digits. Thus we get a special case of the previous question with $k_1 = 9$ and $k_2 = k_3 = \cdots = k_n = 10$. Thus the answer is $9 \cdot 10^{n-1}$ (cf. Exercise 1.3.4).

1.5.1 Draw a tree illustrating the way we counted the number of strings of length 2 formed from the characters a, b, and c, and explain how it gives the answer. Do the same for the more general problem when $n = 3$, $k_1 = 2$, $k_2 = 3$, $k_3 = 2$.

1.5.2 In a sports shop there are T-shirts of 5 different colors, shorts of 4 different colors, and socks of 3 different colors. How many different uniforms can you compose from these items?

1.5.3 On a Toto (soccer poll) ticket, you have to bet 1, 2, or X for each of 13 games. In how many different ways can you fill out the ticket?

1.5.4 We roll a die twice; how many different outcomes can we have? (A 1 followed by a 4 is different from a 4 followed by a 1.)

1.5.5 We have 20 different presents that we want to distribute to 12 children. It is not required that every child get something; it could even happen that we give all the presents to the same child. In how many ways can we distribute the presents?

1.5.6 We have 20 kinds of presents; this time, we have a large supply of each kind. We want to give presents to 12 children. Again, it is not required that every child gets something; but no child can get two copies of the same present. In how many ways can we give presents?

1.6   Permutations

During Alice's birthday party, we encountered the problem of how many ways can we seat n people on n chairs (well, we have encountered it for $n = 6$ and $n = 7$, but the question is natural enough for any n). If we imagine that the seats are numbered, then finding a seating for these people is the same as assigning them to the numbers $1, 2, \ldots, n$ (or $0, 1, \ldots, n-1$ if we want to please the logicians). Yet another way of saying this is to order the people in a single line, or write down an (ordered) list of their names.

If we have a list of n objects (an ordered set, where it is specified which element is the first, second, etc.), and we rearrange them so that they are in another order, this is called *permuting* them; the new order is called a *permutation* of the objects. We also call the rearrangement that does not change anything a permutation (somewhat in the spirit of calling the empty set a set).

For example, the set $\{a, b, c\}$ has the following 6 permutations:

$$abc, acb, bac, bca, cab, cba.$$

So the question is to determine the number of ways n objects can be ordered, i.e., the number of permutations of n objects. The solution found by the people at the party works in general: We can put any of the n people in the first place; no matter whom we choose, we have $n - 1$ choices for the second. So the number of ways to fill the first two positions is $n(n-1)$. No matter how we have filled the first and second positions, there are $n - 2$ choices for the third position, so the number of ways to fill the first three positions is $n(n-1)(n-2)$.

It is clear that this argument goes on like this until all positions are filled. The second to last position can be filled in two ways; the person put in the last position is determined, once the other positions are filled. Thus the number of ways to fill all positions is $n \cdot (n-1) \cdot (n-2) \cdots 2 \cdot 1$. This product is so important that we have a notation for it: $n!$ (read n factorial). In other words, $n!$ is the number of ways to order n objects. With this notation, we can state our second theorem.

Theorem 1.6.1 *The number of permutations of n objects is $n!$.*

Again, we can illustrate the argument above graphically (Figure 1.3). We start with the node on the top, which poses our first decision: Whom do we seat in the first chair? The 3 arrows going out correspond to the three

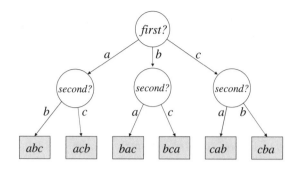

FIGURE 1.3. A decision tree for selecting a permutation of $\{a, b, c\}$.

possible answers to the question. Making a decision, we can follow one of the arrows down to the next node. This carries the next decision problem: whom do we put in the second chair? The two arrows out of the node represent the two possible choices. (Note that these choices are different for different nodes on this level; what is important is that there are two arrows going out from each node.) If we make a decision and follow the corresponding arrow to the next node, we know who sits in the third chair. The node carries the whole seating order.

It is clear that for a set with n elements, n arrows leave the top node, and hence there are n nodes on the next level. Then $n-1$ arrows leave each of these, hence there are $n(n-1)$ nodes on the third level. Then $n-2$ arrows leave each of these, etc. The bottom level has $n!$ nodes. This shows that there are exactly $n!$ permutations.

1.6.1 n boys and n girls go out to dance. In how many ways can they all dance simultaneously? (We assume that only couples of mixed gender dance with each other.)

1.6.2 (a) In how many ways can 8 people play chess in Alice's interpretation of the question?

(b) Can you give a general formula for $2n$ people?

1.7   The Number of Ordered Subsets

At a competition of 100 athletes, only the order of the first 10 is recorded. How many different outcomes does the competition have?

This question can be answered along the lines of the arguments we have seen. The first place can be won by any of the athletes; no matter who wins, there are 99 possible second place winners, so the first two prizes can go $100 \cdot 99$ ways. Given the first two, there are 98 athletes who can be third, etc. So the answer is $100 \cdot 99 \cdots 91$.

1.7.1 Illustrate this argument by a tree.

1.7.2 Suppose that we record the order of all 100 athletes.

(a) How many different outcomes can we have then?

(b) How many of these give the same result for the first 10 places?

(c) Show that the result above for the number of possible outcomes for the first 10 places can be also obtained using (a) and (b).

There is nothing special about the numbers 100 and 10 in the problem above; we could carry out the same for n athletes with the first k places recorded.

To give a more mathematical form to the result, we can replace the athletes by any set of size n. The list of the first k places is given by a sequence of k elements chosen from among n elements, which all have to be different. We may also view this as selecting a subset of the athletes containing k elements, and then ordering them. Thus we have the following theorem.

Theorem 1.7.1 *The number of ordered k-element subsets of a set with n elements is $n(n-1)\cdots(n-k+1)$.*

(Note that if we start with n and count down k numbers, the last one will be $n - k + 1$.)

It is lengthy to talk about "sets with n elements" and "subsets with k elements"; it is convenient to abbreviate these expressions to "n-sets" and "k-subsets." So the number of ordered k-subsets of an n-set is $n(n-1)\cdots(n-k+1)$.

1.7.3 If you generalize the solution of Exercise 1.7.2, you get the answer in the form

$$\frac{n!}{(n-k)!}$$

Check that this is the same number as given in Theorem 1.7.1.

1.7.4 Explain the similarity and the difference between the counting questions answered by Theorem 1.7.1 and Theorem 1.5.1.

1.8   The Number of Subsets of a Given Size

From here, we can easily derive one of the most important counting results.

Theorem 1.8.1 *The number of k-subsets of an n-set is*

$$\frac{n(n-1)\cdots(n-k+1)}{k!} = \frac{n!}{k!(n-k)!}.$$

Proof. Recall that if we count *ordered* subsets, we get $n(n-1)\cdots(n-k+1) = n!/(n-k)!$, by Theorem 1.7.1. Of course, if we want to know the number of *unordered* subsets, then we have overcounted; every subset was counted exactly $k!$ times (with every possible ordering of its elements). So we have to divide this number by $k!$ to get the number of subsets with k elements (without ordering).   □

The number of k-subsets of an n-set is such an important quantity that there is a special notation for it: $\binom{n}{k}$ (read "n choose k"). Thus

$$\binom{n}{k} = \frac{n!}{k!(n-k)!}. \tag{1.6}$$

The number of different lottery tickets is $\binom{90}{5}$, the number of handshakes at the start of Alice's birthday party is $\binom{7}{2}$, etc. The numbers $\binom{n}{k}$ are also called *binomial coefficients* (in Section 3.1 we will see why).

The value of $\binom{n}{n}$ is 1, since an n-element set has exactly one n-element subset, namely itself. It may look a bit more tricky to find that $\binom{n}{0} = 1$, but it is just as easy to explain: Every set has a single 0-element subset, namely the empty set. This is true even for the empty set, so that $\binom{0}{0} = 1$.

1.8.1 Which problems discussed during the party were special cases of theorem 1.8.1?

1.8.2 Tabulate the values of $\binom{n}{k}$ for $0 \leq k \leq n \leq 5$.

1.8.3 Find the values of $\binom{n}{k}$ for $k = 0, 1, n-1, n$ using (1.6), and explain the results in terms of the combinatorial meaning of $\binom{n}{k}$.

Binomial coefficients satisfy many important identities. In the next theorem we collect some of these; some other identities will occur in the exercises and in the next chapter.

Theorem 1.8.2 *Binomial coefficients satisfy the following identities:*

$$\binom{n}{k} = \binom{n}{n-k}; \tag{1.7}$$

If $n, k > 0$, then

$$\binom{n-1}{k-1} + \binom{n-1}{k} = \binom{n}{k}; \tag{1.8}$$

$$\binom{n}{0} + \binom{n}{1} + \binom{n}{2} + \cdots + \binom{n}{n-1} + \binom{n}{n} = 2^n. \tag{1.9}$$

Proof. We prove (1.7) by appealing to the combinatorial meaning of both sides. We have an n-element set, say S. The left hand side counts k-element subsets of S, while the right hand side counts $(n - k)$-element subsets of S. To see that these numbers are the same, we only need to notice that for every k-element subset there is a corresponding $(n - k)$-element subset: its *complement in S* , which contains exactly those elements of S that are not contained in the k-element set. This pairs up the k-element subsets with the $(n-k)$-element subsets, showing that there is the same number of each.

Let's prove (1.8) using the algebraic formula (1.6). After substitution, the identity becomes

$$\frac{n!}{k!(n-k)!} = \frac{(n-1)!}{(k-1)!(n-k)!} + \frac{(n-1)!}{k!(n-k-1)!}.$$

We can divide both sides by $(n-1)!$ and multiply by $(k-1)!(n-k-1)!$; then the identity becomes

$$\frac{n}{k(n-k)} = \frac{1}{n-k} + \frac{1}{k},$$

which can be verified by an easy algebraic computation.

Finally, we prove (1.9) through the combinatorial interpretation again. Again, let S be an n-element set. The first term on the left-hand side counts the 0-element subsets of S (there is only one, the empty set); the second term counts 1-element subsets; the next term, 2-element subsets, etc. In the whole sum, every subset of S is counted exactly once. We know that 2^n (the right-hand side) is the number of all subsets of S. This proves (1.9), and completes the proof of Theorem 1.8.2.   □

1.8.4 Find a proof of (1.7) using the algebraic formula for $\binom{n}{k}$ and a proof of (1.8), using the combinatorial meaning of both sides.

1.8.5 Prove that $\binom{n}{2} + \binom{n+1}{2} = n^2$; give two proofs, one using the combinatorial interpretation, and the other using the algebraic formula for the binomial coefficients.

1.8.6 Prove (again in two ways) that $\binom{n}{k} = \frac{n}{k}\binom{n-1}{k-1}$.

1.8.7 Prove (in two ways) that for $0 \le c \le b \le a$,

$$\binom{a}{b}\binom{b}{c} = \binom{a}{a-c}\binom{a-c}{b-c}$$

Review Exercises

1.8.8 In how many ways can you seat 12 people at two round tables with 6 places each? Think of possible ways of defining when two seatings are different, and find the answer for each.

1.8.9 Name sets with cardinality (a) 365, (b) 12, (c) 7, (d) 11.5, (e) 0, (f) 1024.

1.8.10 List all subsets of $\{a, b, c, d, e\}$ containing $\{a, e\}$ but not containing c.

1.8.11 We have not written up all subset relations between various sets of numbers; for example, $\mathbb{Z} \subseteq \mathbb{R}$ is also true. How many such relations can you find between the sets $\emptyset, \mathbb{N}, \mathbb{Z}_+, \mathbb{Z}, \mathbb{Q}, \mathbb{R}$?

1.8.12 What is the intersection of

(a) the set of positive integers whose last digit is 3, and the set of even numbers;

(b) the set of integers divisible by 5 and the set of even integers?

1.8.13 Let $A = \{a, b, c, d, e\}$ and $B = \{c, d, e\}$. List all subsets of A whose intersection with B has 1 element.

1.8.14 Three sets have 5, 10, and 15 elements, respectively. How many elements can their union and their intersection have?

1.8.15 What is the symmetric difference of A and A?

1.8.16 Form the symmetric difference of A and B to get a set C. Form the symmetric difference of A and C. What set do you get?

1.8.17 Let A, B, C be three sets and assume that A is a subset of C. Prove that

$$A \cup (B \cap C) = (A \cup B) \cap C.$$

Show by an example that the condition that A is a subset of C cannot be omitted.

1.8.18 What is the difference $A \setminus B$ if

(a) A is the set of primes and B is the set of odd integers?

(b) A is the set of nonnegative real numbers and B is the set of nonpositive real numbers?

1.8.19 Prove that for any three sets A, B, C,

$$((A \setminus B) \cup (B \setminus A)) \cap C = ((A \cap C) \cup (B \cap C)) \setminus (A \cap B \cap C).$$

1.8.20 Let A be a set and let $\binom{A}{2}$ denote the set of all 2-element subsets of A. Which of the following statements is true?

$$\binom{A \cup B}{2} = \binom{A}{2} \cup \binom{B}{2}; \qquad \binom{A \cup B}{2} \supseteq \binom{A}{2} \cup \binom{B}{2};$$

$$\binom{A \cap B}{2} = \binom{A}{2} \cap \binom{B}{2}; \qquad \binom{A \cap B}{2} \subseteq \binom{A}{2} \cap \binom{B}{2}.$$

1.8.21 Let B be a subset of A, $|A| = n$, $|B| = k$. What is the number of all subsets of A whose intersection with B has 1 element?

1.8.22 Compute the binary form of 25 and 35, and compute their sum in the binary notation. Check the results against adding 25 and 35 in the usual decimal notation and then converting it to binary.

1.8.23 Prove that every positive integer can be written as the sum of different powers of 2. Also prove that for a given number, there is only one way to do so.

1.8.24 How many bits does 10^{100} have if written in base 2?

1.8.25 Starting from Washington, DC, how many ways can you visit 5 of the 50 state capitals and return to Washington?

1.8.26 Find the number of all 20-digit integers in which no two consecutive digits are the same.

1.8.27 Alice has 10 balls (all different). First, she splits them into two piles; then she picks one of the piles with at least two elements, and splits it into two; she repeats this until each pile has only one element.

(a) How many steps does this take?

(b) Show that the number of different ways in which she can carry out this procedure is

$$\binom{10}{2} \cdot \binom{9}{2} \cdots \binom{3}{2} \cdot \binom{2}{2}.$$

[Hint: Imagine the procedure backward.]

1.8.28 You want to send postcards to 12 friends. In the shop there are only 3 kinds of postcards. In how many ways can you send the postcards, if

(a) there is a large number of each kind of postcard, and you want to send one card to each friend;

(b) there is a large number of each kind of postcard, and you are willing to send one or more postcards to each friend (but no one should get two identical cards);

(c) the shop has only 4 of each kind of postcard, and you want to send one card to each friend?

1.8.29 What is the number of ways to color n objects with 3 colors if every color must be used at least once?

1.8.30 Draw a tree for Alice's solution of enumerating the number of ways 6 people can play chess, and explain Alice's argument using the tree.

1.8.31 How many different "words" can you get by rearranging the letters in the word MATHEMATICS?

1.8.32 Find all positive integers a, b, and c for which

$$\binom{a}{b}\binom{b}{c} = 2\binom{a}{c}.$$

1.8.33 Prove that

$$\binom{n}{k} = \binom{n-2}{k} + 2\binom{n-2}{k-1} + \binom{n-2}{k-2}.$$

1.8.34 20 persons are sitting around a table. How many ways can we choose 3 persons, no two of whom are neighbors?

2
Combinatorial Tools

2.1   Induction

It is time to learn one of the most important tools in discrete mathematics. We start with a question:

We add up the first n odd numbers. What do we get?

Perhaps the best way to try to find the answer is to experiment. If we try small values of n, this is what we find:

$$1 = 1$$
$$1 + 3 = 4$$
$$1 + 3 + 5 = 9$$
$$1 + 3 + 5 + 7 = 16$$
$$1 + 3 + 5 + 7 + 9 = 25$$
$$1 + 3 + 5 + 7 + 9 + 11 = 36$$
$$1 + 3 + 5 + 7 + 9 + 11 + 13 = 49$$
$$1 + 3 + 5 + 7 + 9 + 11 + 13 + 15 = 64$$
$$1 + 3 + 5 + 7 + 9 + 11 + 13 + 15 + 17 = 81$$
$$1 + 3 + 5 + 7 + 9 + 11 + 13 + 15 + 17 + 19 = 100$$

It is easy to observe that we get squares; in fact, it seems from these examples that *the sum of the first n odd numbers is n^2*. We have observed this for the first 10 values of n; can we be sure that it is valid for all? Well, I'd say we can be reasonably sure, but not with mathematical certainty. How can we *prove* the assertion?

Consider the sum for a general n. The nth odd number is $2n-1$ (check!), so we want to prove that

$$1 + 3 + \cdots + (2n - 3) + (2n - 1) = n^2. \tag{2.1}$$

If we separate the last term in this sum, we are left with the sum of the first $(n-1)$ odd numbers:

$$1 + 3 + \cdots + (2n - 3) + (2n - 1) = \left(1 + 3 + \cdots + (2n - 3)\right) + (2n - 1).$$

Now, here the sum in the large parenthesis is $(n-1)^2$, since it is the sum of the first $n-1$ odd numbers. So the total is

$$(n - 1)^2 + (2n - 1) = (n^2 - 2n + 1) + (2n - 1) = n^2, \tag{2.2}$$

just as we wanted to prove.

Wait a minute! Aren't we using in the proof the statement that we are trying to prove? Surely this is unfair! One could prove everything if this were allowed!

In fact, we are not quite using the assertion we are trying to prove. What we were using, was the assertion about the sum of the first $n-1$ odd numbers; and we argued (in (2.2)) that this proves the assertion about the sum of the first n odd numbers. In other words, what we have actually shown is that if the assertion is true for a certain value $(n-1)$, then it is also true for the next value (n).

This is enough to conclude that the assertion is true for every n. We have seen that it is true for $n = 1$; hence by the above, it is also true for $n = 2$ (we have seen this anyway by direct computation, but this shows that this was not even necessary: It follows from the case $n = 1$). In a similar way, the truth of the assertion for $n = 2$ implies that it is also true for $n = 3$, which in turn implies that it is true for $n = 4$, etc. If we repeat this sufficiently many times, we get the truth for any value of n. So the assertion is true for *every* value of n.

This proof technique is called *induction* (or sometimes *mathematical induction*, to distinguish it from a notion in philosophy). It can be summarized as follows.

Suppose that we want to prove a property of positive integers. Also suppose that we can prove two facts:

(a) 1 has the property, and

(b) whenever $n-1$ has the property, then n also has the property $(n > 1)$.

The *Principle of Induction* says that if (a) and (b) are true, then every natural number has the property.

This is precisely what we did above. We showed that the "sum" of the first 1 odd numbers is 1^2, and then we showed that *if* the sum of the first $n-1$ odd numbers is $(n-1)^2$, *then* the sum of the first n odd numbers is n^2, for whichever integer $n > 1$ we consider. Therefore, by the Principle of Induction we can conclude that for every positive integer n, the sum of the first n odd numbers is n^2.

Often, the best way to try to carry out an induction proof is the following. First we prove the statement for $n = 1$. (This is sometimes called the *base case.*) We then try to prove the statement for a general value of n, and we are allowed to assume that the statement is true if n is replaced by $n-1$. (This is called the *induction hypothesis.*) If it helps, one may also use the validity of the statement for $n-2$, $n-3$, etc., and in general, for every k such that $k < n$.

Sometimes we say that if 1 has the property, and every integer n *inherits* the property from $n-1$, then every positive integer has the property. (Just as if the founding father of a family has a certain piece of property, and every new generation inherits this property from the previous generation, then the family will always have this property.)

Sometimes we start not with $n = 1$ but with $n = 0$ (if this makes sense) or perhaps with a larger value of n (if, say, $n = 1$ makes no sense for some reason, or the statement is not for $n = 1$). For example, we want to prove that $n!$ *is an even number if* ≥ 1. We check that this is true for $n = 2$ (indeed, $2! = 2$ is even), and also that it is inherited from $n-1$ to n (indeed, if $(n-1)!$ is even, then so is $n! = n \cdot (n-1)!$, since every multiple of an even number is even). This proves that $n!$ is even for all values of n *from the base case $n = 2$ on.* The assertion is false for $n = 1$, of course.

2.1.1 Prove, using induction but also without it, that $n(n+1)$ is an even number for every nonnegative integer n.

2.1.2 Prove by induction that the sum of the first n positive integers is $n(n+1)/2$.

2.1.3 Observe that the number $n(n+1)/2$ is the number of handshakes among $n+1$ people. Suppose that everyone counts only handshakes with people older than him/her (pretty snobbish, isn't it?). Who will count the largest number of handshakes? How many people count 6 handshakes? (We assume that no two people have exactly the same age.)

Give a proof of the result of Exercise 2.1.2, based on your answer to these questions.

2.1.4 Give a proof of Exercise 2.1.2, based on Figure 2.1.

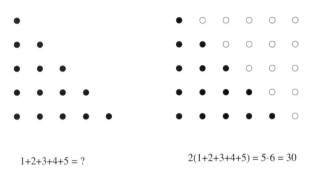

$$1+2+3+4+5 = ? \qquad\qquad 2(1+2+3+4+5) = 5\cdot 6 = 30$$

FIGURE 2.1. The sum of the first n integers

2.1.5 Prove the following identity:

$$1\cdot 2 + 2\cdot 3 + 3\cdot 4 + \cdots + (n-1)\cdot n = \frac{(n-1)\cdot n\cdot (n+1)}{3}.$$

Exercise 2.1.2 relates to a well-known anecdote from the history of mathematics. Carl Friedrich Gauss (1777–1855), one of the greatest mathematicians of all times, was in elementary school when his teacher gave the class the task to add up the integers from 1 to 1000. He was hoping that he would get an hour or so to relax while his students were working. (The story is apocryphal, and it appears with various numbers to add: from 1 to 100, or 1900 to 2000.) To the teacher's great surprise, Gauss came up with the correct answer almost immediately. His solution was extremely simple: Combining the first term with the last, you get $1 + 1000 = 1001$; combining the second term with the last but one, you get $2 + 999 = 1001$; proceeding in a similar way, combining the first remaining term with the last one (and then discarding them) you get 1001. The last pair added this way is $500 + 501 = 1001$. So we obtained 500 times 1001, which makes 500500. We can check this answer against the formula given in exercise 2.1.2: $1000 \cdot 1001/2 = 500500$.

2.1.6 Use the method of the young Gauss to give a third proof of the formula in exercise 2.1.2

2.1.7 How would the young Gauss prove the formula (2.1) for the sum of the first n odd numbers?

2.1.8 Prove that the sum of the first n squares $\left(1 + 4 + 9 + \cdots + n^2\right)$ is $n(n+1)(2n+1)/6$.

2.1.9 Prove that the sum of the first n powers of 2 (starting with $1 = 2^0$) is $2^n - 1$.

In Chapter 1 we often relied on the convenience of saying "etc.": we described some argument that had to be repeated n times to give the result we wanted to get, but after giving the argument once or twice, we said "etc." instead of further repetition. There is nothing wrong with this, if the argument is sufficiently simple so that we can intuitively see where the repetition leads. But it would be nice to have some tool at hand that could be used instead of "etc." in cases where the outcome of the repetition is not so transparent.

The precise way of doing this is using induction, as we are going to illustrate by revisiting some of our results. First, let us give a proof of the formula for the number of subsets of an n-element set, given in Theorem 1.3.1 (recall that the answer is 2^n).

As the Principle of Induction tells us, we have to check that the assertion is true for $n = 0$. This is trivial, and we already did it. Next, we assume that $n > 0$, and that the assertion is true for sets with $n - 1$ elements. Consider a set S with n elements, and fix any element $a \in S$. We want to count the subsets of S. Let us divide them into two classes: those containing a and those not containing a. We count them separately.

First, we deal with those subsets that don't contain a. If we delete a from S, we are left with a set S' with $n - 1$ elements, and the subsets we are interested in are exactly the subsets of S'. By the induction hypothesis, the number of such subsets is 2^{n-1}.

Second, we consider subsets containing a. The key observation is that every such subset consists of a and a subset of S'. Conversely, if we take any subset of S', we can add a to it to get a subset of S containing a. Hence the number of subsets of S containing a is the same as the number of subsets of S', which is, by the induction hypothesis, 2^{n-1}. (We can exercise another bit of mathematical jargon introduced before: The last piece of the argument establishes a one-to-one correspondence between those subsets of S containing a and those not containing a.)

To conclude: The total number of subsets of S is $2^{n-1}+2^{n-1} = 2 \cdot 2^{n-1} = 2^n$. This proves Theorem 1.3.1 (again).

2.1.10 Use induction to prove Theorem 1.5.1 (the number of strings of length n composed of k given elements is k^n) and Theorem 1.6.1 (the number of permutations of a set with n elements is $n!$).

2.1.11 Use induction on n to prove the "handshake theorem" (the number of handshakes between n people is $n(n-1)/2$).

2.1.12 Read carefully the following induction proof:

ASSERTION: $n(n+1)$ is an odd number for every n.

PROOF: Suppose that this is true for $n - 1$ in place of n; we prove it for n, using the induction hypothesis. We have

$$n(n+1) = (n-1)n + 2n.$$

Now here $(n-1)n$ is odd by the induction hypothesis, and $2n$ is even. Hence $n(n + 1)$ is the sum of an odd number and an even number, which is odd.

The assertion that we proved is obviously wrong for $n = 10$: $10 \cdot 11 = 110$ is even. What is wrong with the proof?

2.1.13 Read carefully the following induction proof:

ASSERTION: *If we have n lines in the plane, no two of which are parallel, then they all go through one point.*

PROOF: The assertion is true for one line (and also for 2, since we have assumed that no two lines are parallel). Suppose that it is true for any set of $n - 1$ lines. We are going to prove that it is also true for n lines, using this induction hypothesis.

So consider a set $S = \{a, b, c, d, \dots\}$ of n lines in the plane, no two of which are parallel. Delete the line c; then we are left with a set S' of $n - 1$ lines, and obviously, no two of these are parallel. So we can apply the induction hypothesis and conclude that there is a point P such that all the lines in S' go through P. In particular, a and b go through P, and so P must be the point of intersection of a and b.

Now put c back and delete d, to get a set S'' of $n - 1$ lines. Just as above, we can use the induction hypothesis to conclude that these lines go through the same point P'; but just as above, P' must be the point of intersection of a and b. Thus $P' = P$. But then we see that c goes through P. The other lines also go through P (by the choice of P), and so all the n lines go through P.

But the assertion we proved is clearly wrong; where is the error?

2.2   Comparing and Estimating Numbers

It is nice to have formulas for certain numbers (for example, for the number $n!$ of permutations of n elements), but it is often more important to have a rough idea about how large these numbers are. For example, how many digits does 100! have?

Let us start with simpler questions. Which is larger, n or $\binom{n}{2}$? For $n = 2, 3, 4$ the value of $\binom{n}{2}$ is $1, 3, 6$, so it is less than n for $n = 2$, equal for $n = 3$, but larger for $n = 4$. In fact, $n = \binom{n}{1} < \binom{n}{2}$ if $n \geq 4$.

More can be said: The quotient

$$\frac{\binom{n}{2}}{n} = \frac{n - 1}{2}$$

becomes arbitrarily large as n becomes large; for example, if we want this quotient to be larger than 1000, it suffices to choose $n > 2001$. In the language of calculus, we have

$$\frac{\binom{n}{2}}{n} \to \infty \qquad (n \to \infty).$$

Here is another simple question: Which is larger, n^2 or 2^n? For small values of n, this can go either way: $1^2 < 2^1$, $2^2 = 2^2$, $3^2 > 2^3$, $4^2 = 2^4$, $5^2 < 2^5$. But from here on, 2^n takes off and grows much faster than n^2. For example, $2^{10} = 1024$ is much larger than $10^2 = 100$. In fact, $2^n/n^2$ becomes arbitrarily large, as n becomes large.

2.2.1   (a) Prove that $2^n > \binom{n}{3}$ if $n \geq 3$.

(b) Use (a) to prove that $2^n/n^2$ becomes arbitrarily large as n becomes large.

Now we tackle the problem of estimating $100!$ or, more generally, $n! = 1 \cdot 2 \cdots n$. The first factor 1 does not matter, but all the others are at least 2, so $n! \geq 2^{n-1}$. Similarly, $n! \leq n^{n-1}$, since (ignoring the factor 1 again) $n!$ is the product of $n - 1$ factors, each of which is at most n. (Since all but one of them are smaller than n, the product is in fact much smaller.) Thus we know that

$$2^{n-1} \leq n! \leq n^{n-1}. \tag{2.3}$$

These bounds are very far apart; for $n = 10$, the lower bound is $2^9 = 512$, while the upper bound is 10^9 (one billion).

Here is a question that is not answered by the simple bounds in (2.3). Which is larger, $n!$ or 2^n? In other words, does a set with n elements have more permutations or more subsets? For small values of n, subsets are winning: $2^1 = 2 > 1! = 1$, $2^2 = 4 > 2! = 2$, $2^3 = 8 > 3! = 6$. But then the picture changes: $2^4 = 16 < 4! = 24$, $2^5 = 32 < 5! = 120$. It is easy to see that as n increases, $n!$ grows much faster than 2^n: If we go from n to $n+1$, then 2^n grows by a factor of 2, while $n!$ grows by a factor of $n + 1$.

2.2.2 Use induction to make the previous argument precise, and prove that $n! > 2^n$ if $n \geq 4$.

There is a formula that gives a very good approximation of $n!$. We state it without proof, since the proof (although not terribly difficult) needs calculus.

Theorem 2.2.1 [Stirling's formula]

$$n! \sim \left(\frac{n}{e}\right)^n \sqrt{2\pi n}.$$

Here $\pi = 3.14\ldots$ is the area of the circle with unit radius, $e = 2.718\ldots$ is the base of the natural logarithm, and $\sim$ means approximate equality in the precise sense that

$$\frac{n!}{\left(\frac{n}{e}\right)^n \sqrt{2\pi n}} \to 1 \qquad (n \to \infty).$$

Both of the funny irrational numbers e and π occur in the same formula!

Let us return to the question: How many digits does 100! have? We know by Stirling's formula that

$$100! \approx (100/e)^{100} \cdot \sqrt{200\pi}.$$

The number of digits of this number is its logarithm, in base 10, rounded up. Thus we get

$$\lg(100!) \approx 100\lg(100/e) + 1 + \lg\sqrt{2\pi} = 157.969\ldots.$$

So the number of digits in 100! is about 158 (this is, in fact, the right value).

2.3   Inclusion-Exclusion

In a class of 40, many students are collecting the pictures of their favorite rock stars. Eighteen students have a picture of the Beatles, 16 students have a picture of the Rolling Stones and 12 students have a picture of Elvis Presley (this happened a long time ago, when we were young). There are 7 students who have pictures of both the Beatles and the Rolling Stones, 5 students who have pictures of both the Beatles and Elvis Presley, and 3 students who have pictures of both the Rolling Stones and Elvis Presley. Finally, there are 2 students who possess pictures of all three groups. Question: How many students in the class have no picture of any of the rock groups?

First, we may try to argue like this: There are 40 students altogether in the class; take away from this the number of those having Beatles pictures (18), those having Rolling Stones picture (16), and those having Elvis pictures (12); so we take away $18 + 16 + 12$. We get -6; this negative number warns us that there must be some error in our calculation; but what was not correct? We made a mistake when we subtracted the number of those students who collected the pictures of two groups twice! For example, a student having the Beatles and Elvis Presley was subtracted with the Beatles collectors as well as with the Elvis Presley collectors. To correct our calculations, we have to add back the number of those students who have pictures of two groups. This way we get $40 - (18 + 16 + 12) + (7 + 5 + 3)$. But we must be careful; we shouldn't make the same mistake again! What happened to the 2 students who have the pictures of all three groups? We subtracted these 3 times at the beginning, and then we added them back 3 times, so we must subtract them once more! With this correction, our final result is:

$$40 - (18 + 16 + 12) + (7 + 5 + 3) - 2 = 7. \tag{2.4}$$

We can not find any error in this formula, looking at it from any direction. But learning from our previous experience, we must be much more careful: We have to give an exact proof!

So suppose that somebody records picture collecting data of the class in a table like Table 2.1 below. Each row corresponds to a student; we did not put down all the 40 rows, just a few typical ones.

Name	Bonus	Beatles	Stones	Elvis	BS	BE	SE	BSE
Al	1	0	0	0	0	0	0	0
Bel	1	−1	0	0	0	0	0	0
Cy	1	−1	−1	0	1	0	0	0
Di	1	−1	0	−1	0	1	0	0
Ed	1	−1	−1	−1	1	1	1	−1
⋮								

TABLE 2.1. Strange record of who's collecting whose pictures.

The table is a bit silly (but with reason). First, we give a bonus of 1 to every student. Second, we record in a separate column whether the student is collecting (say) both the Beatles and Elvis Presley (the column labeled BE), even though this could be read off from the previous columns. Third, we put a −1 in columns recording the collecting of an odd number of pictures, and a 1 in columns recording the collecting of an even number of pictures.

We compute the total sum of entries in this table in two different ways. First, what are the row sums? We get 1 for Al and 0 for everybody else. This is not a coincidence. If we consider a student like Al, who does not have any picture, then this student contributes to the bonus column, but nowhere else, which means that the sum in the row of this student is 1. Next, consider Ed, who has all 3 pictures. He has a 1 in the bonus column; in the next 3 columns he has 3 terms that are −1. In each of the next 3 columns he has a 1, one for each pair of pictures; it is better to think of this 3 as $\binom{3}{2}$. His row ends with $\binom{3}{3}$ −1's ($\binom{3}{3}$ equals 1, but in writing it this way the general idea can be seen better). So the sum of the row is

$$1 - \binom{3}{1} + \binom{3}{2} - \binom{3}{3} = 0.$$

Looking at the rows of Bel, Cy, and Di, we see that their sums are

$$1 - \binom{1}{1} = 0 \qquad \text{for Bel (1 picture),}$$

$$1 - \binom{2}{1} + \binom{2}{2} = 0 \qquad \text{for Cy and Di (2 pictures).}$$

If we move the negative terms to the other side of these equations, we get an equation with a combinatorial meaning: *The number of subsets of an n-set with an even number of elements is the same as the number of subsets with an odd number of elements.* For example,

$$\binom{3}{0} + \binom{3}{2} = \binom{3}{1} + \binom{3}{3}.$$

Recall that Exercise 1.3.3 asserts that this is indeed so for every $n \geq 1$.

Since the row sum is 0 for all those students who have any picture of any music group, and it is 1 for those having no picture at all, the sum of all 40 row sums gives exactly the number of those students having no picture at all.

On the other hand, what are the column sums? In the "bonus" column, we have 40 times $+1$; in the "Beatles" column, we have 18 times -1; then we have 16 and 12 times -1. Furthermore, we get 7 times $+1$ in the BS column, then 5- and 3 times $+1$ in the BE and SE columns, respectively. Finally, we get 2 -1's in the last column. So this is indeed the expression in (2.4).

This formula is called the *Inclusion–Exclusion Formula* or *Sieve Formula*. The origin of the first name is obvious; the second refers to the image that we start with a large set of objects and then "sieve out" those objects we don't want to count.

We could extend this method if students were collecting pictures of 4, or 5, or any number of rock groups instead of 3. Rather than stating a general theorem (which would be lengthy), we give a number of exercises and examples.

2.3.1 In a class of all boys, 18 boys like to play chess, 23 like to play soccer, 21 like biking and 17 like hiking. The number of those who like to play both chess and soccer is 9. We also know that 7 boys like chess and biking, 6 boys like chess and hiking, 12 like soccer and biking, 9 boys like soccer and hiking, and finally 12 boys like biking and hiking. There are 4 boys who like chess, soccer, and biking, 3 who like chess, soccer, and hiking, 5 who like chess, biking, and hiking, and 7 who like soccer, biking, and hiking. Finally there are 3 boys who like all four activities. In addition we know that everybody likes at least one of these activities. How many boys are there in the class?

2.4   Pigeonholes

Can we find in New York two persons having the same number of strands of hair? One would think that it is impossible to answer this question, since one does not even know how many strands of hair there are on one's own head, let alone about the number of strands of hair on every person living in New York (whose exact number is in itself quite difficult to determine). But there are some facts that we know for sure: Nobody has more than 500,000 strands of hair (a scientific observation), and there are more than 10 million inhabitants of New York. Can we now answer our original question? Yes. If there were no two people with the same number of strands of hair, then there would be at most one person having 0 strands, at most one person having exactly 1 strand, and so on. Finally, there would be at most one person having exactly 500,000 strands. But then this means that there are no more than 500,001 inhabitants of New York. Since this contradicts what we know about New York, it follows that there must be two people having the same number of strands of hair.[1]

We can formulate our solution as follows. Imagine 500,001 enormous boxes (or pigeon holes). The first one is labeled "New Yorkers having 0 strands of hair," the next is labeled "New Yorkers having 1 strand of hair," and so on. The last box is labeled "New Yorkers having 500,000 strands of hair". Now if everybody goes to the proper box, then about 10 million New Yorkers are properly assigned to some box (or hole). Since we have only 500,001 boxes, there certainly will be a box containing more than one New Yorker. This statement is obvious, but it is very often a powerful tool, so we formulate it in full generality:

> *If we have n boxes and we place more than n objects into them, then there will be at least one box that contains more than one object.*

Very often, the above statement is formulated using pigeons and their holes, and is referred to as the *Pigeonhole Principle*. The Pigeonhole Principle is simple indeed: Everybody understands it immediately. Nevertheless, it deserves a name, since we use it very often as the basic tool of many proofs. We will see many examples for the use of the Pigeonhole Principle, but to show you its power, we discuss one of them right away. This is not a theorem of any significance; rather, and exercise whose solution is given in detail.

Exercise. *We shoot 50 shots at a square target, the side of which is 70 cm long. We are quite a good shot, because all of our shots hit the target. Prove that there are two bulletholes that are closer than 15 cm.*

Solution: Imagine that our target is an old chessboard. One row and one column of it has fallen off, so it has 49 squares. The board received 50 shots, so there must be a square that received at least two shots (putting bulletholes in pigeonholes). We claim that these two shots are closer to each other than 15 cm.

---

[1]There is an interesting feature of this argument: we end up knowing that two such persons exist, without having the slightest hint about how to find these people. (Even if we suspect that two people have the same number of strands of hair, it is essentially impossible to verify that this is indeed so!) Such proofs in mathematics are called *pure existence proofs*.

The side of the square is obviously 10 cm, the diagonals of it are equal, and (from the Pythagorean Theorem) their length is $\sqrt{200} \approx 14.1$ cm. We show that

(∗) *the two shots cannot be at a larger distance than the diagonal.*

It is intuitively clear that two points in the square at largest distance are the endpoints of one of the diagonals, but intuition can be misleading; let us prove this. Suppose that two points P and Q are farther away then the length of the diagonal. Let A, B, C, and D be the vertices of the square. Connect P and Q by a line, and let P' and Q' be the two points where this line intersects the boundary of the square (Figure 2.2). Then the distance of P' and Q' is even larger, so it is also larger than the diagonal.

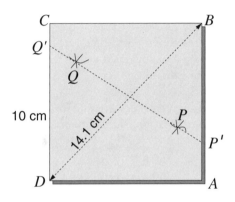

FIGURE 2.2. Two shots in the same square

We may assume without loss of generality that P' lies on the side AB (if this is not the case, we change the names of the vertices). One of the angles $Q'P'A$ and $Q'P'B$ is at least 90°; we may assume (again without loss of generality) that $Q'P'A$ is this angle. Then the segment AQ' is the edge of the triangle $Q'P'A$ opposite the largest angle, and so it is even longer than $P'Q'$, and so it is longer than the diagonal.

We repeat this argument to show that if we replace Q' by one of the endpoints of the side it lies on, we get a segment that is longer than the diagonal. But now we have a segment both of whose endpoints are vertices of the square. So this segment is either a side or a diagonal of the square, and in neither case is it longer than the diagonal! This contradiction shows that the assertion (∗) above must be true.

So we got not only that there will be two shots that are closer than 15 cm, but even closer than 14.2 cm. This concludes the solution of the exercise.

If this is the first time you have seen this type of proof, you may be surprised: we did not argue directly to prove what we wanted, but instead assumed that the assertion was not true, and then using this additional assumptions, we argued until we got a contradiction. This form of proof is called *indirect*, and it is quite often used in mathematical reasoning, as we will see throughout this book. (Mathematicians are strange creatures, one may observe: They go into long arguments based on assumptions they know are false, and their happiest moments are when they find a contradiction between statements they have proved.)

2.4.1 Prove that we can select 20 New Yorkers who all have the same number of strands of hair.

2.5   The Twin Paradox and the Good Old Logarithm

Having taught the Pigeonhole Principle to his class, the professor decides to play a little game: "I bet that there are two of you who have the same birthday! What do you think?" Several students reply immediately: "There are 366 possible birthdays, so you could only conclude this if there were at least 367 of us in the class! But there are only 50 of us, and so you'd lose the bet." Nevertheless, the professor insists on betting, and he wins.

How can we explain this? The first thing to realize is that the Pigeonhole Principle tells us that with 367 students in the class, the professor *always* wins the bet. But this is uninteresting as bets go; it is enough for him that he has a good chance of winning. With 366 students, he may already lose; could it be that with only 50 students he still has a good chance of winning?

The surprising answer is that even with as few as 23 students, his chance of winning is slightly larger than 50%. We can view this fact as a "Probabilistic Pigeonhole Principle", but the usual name for it is the *Twin Paradox*.

Let us try to determine the professor's chances. Suppose that on the class list, he writes down everybody's birthday. So he has a list of 50 birthdays. We know from Section 1.5 that there are 366^{50} different lists of this type.

For how many of these does he lose? Again, we already know the answer from Section 1.7: $366 \cdot 365 \cdots 317$. So the probability that he loses the bet is[2]

$$\frac{366 \cdot 365 \cdots 317}{366^{50}}.$$

With some effort, we could calculate this value "by brute force", using a computer (or just a programmable calculator), but it will be much more

---

[2]Here we made the implicit assumption that all the 366^{50} birthday lists are equally likely. This is certainly not true; for example, lists containing February 29 are clearly much less likely. There are also (much smaller) variations between the other days of the year. It can be shown, however, that these variations only help the professor, making collisions of birthdays more likely.

useful to get upper and lower bounds by a method that will work in a more general case, when we have n possible birthdays and k students. In other words, how large is the quotient

$$\frac{n(n-1)\cdots(n-k+1)}{n^k} ?$$

It will be more convenient to take the reciprocal (which is then larger than 1):

$$\frac{n^k}{n(n-1)\cdots(n-k+1)}. \tag{2.5}$$

We can simplify this fraction by cancelling an n, but then there is no obvious way to continue. A little clue may be that the number of factors is the same in the numerator and denominator, so let us try to write this fraction as a product:

$$\frac{n^k}{n(n-1)\ldots(n-k+1)} = \frac{n}{n-1}\cdot\frac{n}{n-2}\cdots\frac{n}{n-k+1}.$$

These factors are quite simple, but it is still difficult to see how large their product is. The individual factors are larger than 1, but (at least at the beginning) quite close to 1. But there are many of them, and their product may be large.

The following idea helps: *Take the logarithm!*[3] We get

$$\ln\left(\frac{n^k}{n(n-1)\cdots(n-k+1)}\right) = \ln\left(\frac{n}{n-1}\right) + \ln\left(\frac{n}{n-2}\right) + \cdots$$
$$+ \ln\left(\frac{n}{n-k+1}\right). \tag{2.6}$$

(Naturally, we took the natural logarithm, base $e = 2.71828\ldots$.) This way we can deal with addition instead of multiplication, which is nice; but the terms we have to add up became much uglier! What do we know about these logarithms?

Let's look at the graph of the logarithm function (Figure 2.3). We have also drawn the line $y = x - 1$. We see that the function is below the line, and touches it at the point $x = 1$ (these facts can be proved by really elementary calculus). So we have

$$\ln x \leq x - 1. \tag{2.7}$$

Can we say something about how good this upper bound is? From the figure we see that at least for values of x close to 1, the two graphs are

---

[3] After all, the logarithm was invented in the seventeenth century by Buergi and Napier to make multiplication easier, by turning it into addition.

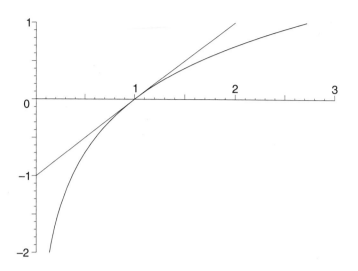

FIGURE 2.3. The graph of the natural logarithm function. Note that near 1, it is very close to the line $x - 1$.

quite close. Indeed, we can do the following little computation:

$$\ln x = -\ln \frac{1}{x} \geq -\left(\frac{1}{x} - 1\right) = \frac{x - 1}{x}. \qquad (2.8)$$

If x is a little larger than 1 (as are the values we have in (2.6)), then $\frac{x-1}{x}$ is only a little smaller than $x - 1$, and so the upper bound in (2.7) and lower bound in (2.8) are quite close.

These bounds on the logarithm function are very useful in many applications in which we have to do approximate computations with logarithms, and it is worthwhile to state them in a separate lemma. (A lemma is a precise mathematical statement, just like a theorem, except that it is not the goal itself, but some auxiliary result used along the way to the proof of a theorem. Of course, some lemmas are more interesting than some theorems!)

Lemma 2.5.1 *For every $x > 0$,*

$$\frac{x - 1}{x} \leq \ln x \leq x - 1.$$

First we use the lower bound in this lemma to estimate (2.6) from below. For a typical term in the sum in (2.6) we get

$$\ln\left(\frac{n}{n - j}\right) \geq \frac{\frac{n}{n-j} - 1}{\frac{n}{n-j}} = \frac{j}{n},$$

and hence

$$\ln\left(\frac{n^k}{n(n-1)\cdots(n-k+1)}\right) \geq \frac{1}{n} + \frac{2}{n} + \cdots + \frac{k-1}{n}$$

$$= \frac{1}{n}(1 + 2 + \cdots + (k-1)) = \frac{k(k-1)}{2n}$$

(remember the young Gauss's problem!). Thus we have a simple lower bound on (2.6). To get an upper bound, we can use the other inequality in Lemma 2.5.1; for a typical term, we get

$$\ln\left(\frac{n}{n-j}\right) \leq \frac{n}{n-j} - 1 = \frac{j}{n-j}.$$

We have to sum these for $j = 1, \ldots, k-1$ to get an upper bound on (2.6). This is not as easy as in young Gauss's case, since the denominator is changing. But we only want an upper bound, so we could replace the denominator by the smallest value it can have for various values of j, namely $n-k+1$. We have $j/(n-j) \leq j/(n-k+1)$, and hence

$$\ln\left(\frac{n^k}{n(n-1)\cdots(n-k+1)}\right) \leq \frac{1}{n-k+1} + \frac{2}{n-k+1} + \cdots + \frac{k-1}{n-k+1}$$

$$= \frac{1}{n-k+1}(1 + 2 + \cdots + (k-1))$$

$$= \frac{k(k-1)}{2(n-k+1)}.$$

Thus we have these similar upper and lower bounds on the logarithm of the ratio (2.5), and applying the exponential function to both sides, we get the following:

$$e^{\frac{k(k-1)}{2n}} \leq \frac{n^k}{n(n-1)\cdots(n-k+1)} \leq e^{\frac{k(k-1)}{2(n-k+1)}}. \tag{2.9}$$

Does this help to understand the professor's trick in the classroom? Let's apply (2.9) with $n = 366$ and $k = 50$; using our calculators, we get that

$$28.4 \leq \frac{366^{50}}{366 \cdot 364 \cdots 317} \leq 47.7.$$

(Using more computation, we can determine that the exact value is $33.414\ldots$.) So the probability that all students in the class have different birthdays (which is the reciprocal of this number) is less than $1/28$. This means that if the professor performs this trick every year, he will likely fail only once or twice in his career!

Review Exercises

2.5.1 What is the following sum?

$$\frac{1}{1\cdot 2} + \frac{1}{2\cdot 3} + \frac{1}{3\cdot 4} + \cdots + \frac{1}{(n-1)\cdot n}.$$

Experiment, conjecture the value, and then prove it by induction.

2.5.2 What is the following sum?

$$0\cdot\binom{n}{0} + 1\cdot\binom{n}{1} + 2\cdot\binom{n}{2} + \cdots + (n-1)\cdot\binom{n}{n-1} + n\cdot\binom{n}{n}.$$

Experiment, conjecture the value, and then prove it. (Try to prove the result by induction and also by combinatorial arguments.)

2.5.3 Prove the following identities:

$$1\cdot 2^0 + 2\cdot 2^1 + 3\cdot 2^2 + \cdots + n\cdot 2^{n-1} = (n-1)2^n + 1.$$

$$1^3 + 2^3 + 3^3 + \cdots + n^3 = \frac{n^2(n+1)^2}{4}.$$

$$1 + 3 + 9 + 27 + \cdots + 3^{n-1} = \frac{3^n - 1}{2}.$$

2.5.4 Prove by induction on n that

(a) $n^2 - 1$ is a multiple of 4 if n is odd,

(b) $n^3 - n$ is a multiple of 6 for every n.

2.5.5 There is a class of 40 girls. There are 18 girls who like to play chess, and 23 who like to play soccer. Several of them like biking. The number of those who like to play both chess and soccer is 9. There are 7 girls who like chess and biking, and 12 who like soccer and biking. There are 4 girls who like all three activities. In addition we know that everybody likes at least one of these activities. How many girls like biking?

2.5.6 There is a class of all boys. We know that there are a boys who like to play chess, b who like to play soccer, c who like biking and d who like hiking. The number of those who like to play both chess and soccer is x. There are y boys who like chess and biking, z boys who like chess and hiking, u who like soccer and biking, v boys who like soccer and hiking, and finally w boys who like biking and hiking. We don't know how many boys like, e.g.,chess, soccer and hiking, but we know that everybody likes at least one of these activities. We would like to know how many boys are in the class.

(a) Show by an example that this is not determined by what we know.

(b) Prove that we can at least conclude that the number of boys in the class is at most $a + b + c + d$, and at least $a + b + c + d - x - y - z - u - v - w$.

2.5.7 We select 38 even positive integers, all less than 1000. Prove that there will be two of them whose difference is at most 26.

2.5.8 A drawer contains 6 pairs of black, 5 pairs of white, 5 pairs of red, and 4 pairs of green socks.

(a) How many single socks do we have to take out to make sure that we take out two socks with the same color?

(b) How many single socks do we have to take out to make sure that we take out two socks with different colors?

3
Binomial Coefficients and Pascal's Triangle

3.1 The Binomial Theorem

In Chapter 1 we introduced the numbers $\binom{n}{k}$ and called them *binomial coefficients*. It is time to explain this strange name: it comes from a very important formula in algebra involving them, which we discuss next.

The issue is to compute powers of the simple algebraic expression $(x+y)$. We start with small examples:

$$(x + y)^2 = x^2 + 2xy + y^2,$$
$$(x + y)^3 = (x + y) \cdot (x + y)^2 = (x + y) \cdot (x^2 + 2xy + y^2)$$
$$= x^3 + 3x^2y + 3xy^2 + y^3,$$

and continuing like this,

$$(x + y)^4 = (x + y) \cdot (x + y)^3 = x^4 + 4x^3y + 6x^2y^2 + 4xy^3 + y^4.$$

These coefficients are familiar! We have seen them, e.g., in exercise 1.8.2, as the numbers $\binom{n}{k}$. Let us make this observation precise. We illustrate the argument for the next value of n, namely $n = 5$, but it works in general.

Think of expanding

$$(x + y)^5 = (x + y)(x + y)(x + y)(x + y)(x + y)$$

so that we get rid of all parentheses. We get each term in the expansion by selecting one of the two terms in each factor, and multiplying them. If we choose x, say, 2 times, then we must choose y 3 times, and so we get x^2y^3. How many times do we get this same term? Clearly, as many times as the number of ways to select the three factors that supply y (the remaining factors supply x). Thus we have to choose three factors out of 5, which can be done in $\binom{5}{3}$ ways.

Hence the expansion of $(x+y)^5$ looks like this:

$$(x+y)^5 = \binom{5}{0}x^5 + \binom{5}{1}x^4y + \binom{5}{2}x^3y^2 + \binom{5}{3}x^2y^3 + \binom{5}{4}xy^4 + \binom{5}{5}y^5.$$

We can apply this argument in general to obtain the *Binomial Theorem*:

Theorem 3.1.1 (The Binomial Theorem) *The coefficient of $x^{n-k}y^k$ in the expansion of $(x+y)^n$ is $\binom{n}{k}$. In other words, we have the identity*

$$(x+y)^n = \binom{n}{0}x^n + \binom{n}{1}x^{n-1}y + \binom{n}{2}x^{n-2}y^2 + \cdots + \binom{n}{n-1}xy^{n-1} + \binom{n}{n}y^n.$$

This important identity was discovered by the famous Persian poet and mathematician Omar Khayyam (1044?–1123?). Its name comes from the Greek word *binome* for an expression consisting of two terms, in this case, $x+y$. The appearance of the numbers $\binom{n}{k}$ in this theorem is the source of their name: *binomial coefficients*.

The Binomial Theorem can be applied in many ways to get identities concerning binomial coefficients. For example, let us substitute $x = y = 1$. Then we get identity (1.9):

$$2^n = \binom{n}{0} + \binom{n}{1} + \binom{n}{2} + \cdots + \binom{n}{n-1} + \binom{n}{n}. \qquad (3.1)$$

Later on, we are going to see trickier applications of this idea. For the time being, another twist on it is contained in exercise (3.1.2).

3.1.1 Give a proof of the Binomial Theorem by induction, based on (1.8).

3.1.2 (a) Prove the identity

$$\binom{n}{0} - \binom{n}{1} + \binom{n}{2} - \binom{n}{3} + \cdots = 0.$$

(The sum ends with $\binom{n}{n} = 1$, with the sign of the last term depending on the parity of n.)

(b) This identity is obvious if n is odd. Why?

3.1.3 Prove the identity in Exercise 3.1.2, using a combinatorial interpretation of the positive and negative terms.

3.2   Distributing Presents

Suppose we have n different presents, which we want to distribute to k children, where for some reason, we are told how many presents each child should get. So Adam should get n_{Adam} presents, Barbara, n_{Barbara} presents, etc. In a mathematically convenient (though not very friendly) way, we call the children $1, 2, \ldots, k$; thus we are given the numbers (nonnegative integers) $n_1, n_2, \ldots, n_k$. We assume that $n_1 + n_2 + \cdots + n_k = n$, else there is no way to distribute all the presents and give each child the right number of them.

The question is, of course, how many ways can these presents be distributed?

We can organize the distribution of presents as follows. We lay out the presents in a single row of length n. The first child comes and takes the first n_1 presents, starting from the left. Then the second comes and takes the next n_2; then the third takes the next n_3 presents etc. Child k gets the last n_k presents.

It is clear that we can determine who gets what by choosing the order in which the presents are laid out. There are $n!$ ways to order the presents. But of course, the number $n!$ overcounts the number of ways to distribute the presents, since many of these orderings lead to the same results (that is, every child gets the same set of presents). The question is, how many?

So let us start with a given distribution of presents, and let's ask the children to lay out the presents for us, nicely in a row, starting with the first child, then continuing with the second, third, etc. This way we get back *one* possible ordering that leads to the current distribution. The first child can lay out his presents in $n_1!$ possible orders; no matter which order he chooses, the second child can lay out her presents in $n_2!$ possible ways, etc. So the number of ways the presents can be laid out (given the distribution of the presents to the children) is a product of factorials:

$$n_1! \cdot n_2! \cdots n_k! .$$

Thus the number of ways of distributing the presents is

$$\frac{n!}{n_1! n_2! \cdots n_k!} .$$

3.2.1 We can describe the procedure of distributing the presents as follows. First, we select n_1 presents and give them to the first child. This can be done in $\binom{n}{n_1}$ ways. Then we select n_2 presents from the remaining $n - n_1$ and give them to the second child, etc.

Complete this argument and show that it leads to the same result as the previous one.

3.2.2 The following special cases should be familiar from previous problems and theorems. Explain why.

(a) $n = k$, $n_1 = n_2 = \cdots = n_k = 1$;

(b) $n_1 = n_2 = \cdots = n_{k-1} = 1$, $n_k = n - k + 1$;

(c) $k = 2$;

(d) $k = 3$, $n = 6$, $n_1 = n_2 = n_3 = 2$.

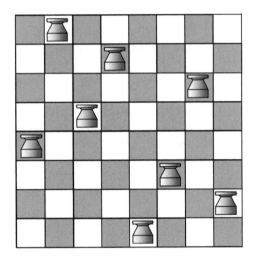

FIGURE 3.1. Placing 8 nonattacking rooks on a chessboard.

3.2.3 (a) How many ways can you place n rooks on a chessboard so that no two attack each other (Figure 3.1)? We assume that the rooks are identical, so interchanging two rooks does not count as a separate placement.

(b) How many ways can you do this if you have 4 wooden and 4 marble rooks?

(c) How many ways can you do this if all the 8 rooks are different?

3.3 Anagrams

Have you played with anagrams? One selects a word (say, COMBINA-TORICS) and tries to compose from its letters meaningful or, even better, funny words or expressions.

How many anagrams can you build from a given word? If you try to answer this question by playing around with the letters, you will realize that the question is badly posed; it is difficult to draw the line between meaningful and nonmeaningful anagrams. For example, it could easily happen that A CROC BIT SIMON. And it may be true that Napoleon always wanted a TOMB IN CORSICA. It is questionable, but certainly grammatically correct, to assert that COB IS ROMANTIC. Some universities may have a course on MAC IN ROBOTICS.

But one would have to write a book to introduce an exciting character, ROBIN COSMICAT, who enforces a COSMIC RIOT BAN, while appealing TO COSMIC BRAIN.

And it would be terribly difficult to explain an anagram like MTBIRAS-CIONOC.

To avoid this controversy, let's accept everything; i.e., we don't require the anagram to be meaningful (or even pronounceable). Of course, the production of anagrams then becomes uninteresting; but at least we can tell how many of them there are!

3.3.1 How many anagrams can you make from the word COMBINATORICS?

3.3.2 Which word gives rise to more anagrams: COMBINATORICS or COMBINATORICA? (The latter is the Latin name of the subject.)

3.3.3 Which word with 13 letters gives rise to the most anagrams? Which word gives rise to the least?

So let's see the general answer to the question of counting anagrams. If you have solved the problems above, it should be clear that the number of anagrams of an n-letter word depends on how many times letters of the word are repeated. So suppose that there are k letters $A, B, C, \ldots Z$ in the alphabet, and the word contains letter A n_1 times (this could be 0), letter B, n_2 times, etc., letter Z, n_k times. Clearly, $n_1 + n_2 + \cdots + n_k = n$.

Now, to form an anagram, we have to select n_1 positions for letter A, n_2 positions for letter B, etc., n_k positions for letter Z. Having formulated it this way, we can see that this is nothing but the question of distributing n presents to k children when it is prescribed how many presents each child gets. Thus we know from the previous section that the answer is

$$\frac{n!}{n_1! n_2! \cdots n_k!}.$$

3.3.4 It is clear that STATUS and LETTER have the same number of anagrams (in fact, $6!/(2!2!) = 180$). We say that these words are "essentially the same" (at least as far as counting anagrams goes): They have two letters repeated twice and two letters occurring only once. We call two words "essentially different", if they are not "essentially the same."

(a) How many 6-letter words are there, if, to begin with, we consider any two words different if they are not completely identical? (As before, the words don't have to be meaningful. The alphabet has 26 letters.)

(b) How many words with 6 letters are "essentially the same" as the word LETTER?

(c) How many "essentially different" 6-letter words are there?

(d) Try to find a general answer to question (c) (that is, how many "essentially different" words are there with n letters?). If you can't find the answer, read the following section and return to this exercise afterwards.

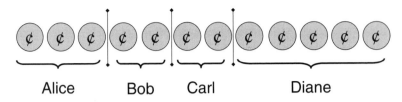

FIGURE 3.2. How to distribute n pennies to k children?

3.4 Distributing Money

Instead of distributing presents, let's distribute money. Let us formulate the question in general: We have n pennies that we want to distribute among k kids. Each child must get at least one penny (and, of course, an integer number of pennies). How many ways can we distribute the money?

Before answering this question, we must clarify the difference between distributing money and distributing presents. If you are distributing presents, you have to decide not only how many presents each child gets, but also *which* of the different presents the child gets. If you are distributing money, only the quantity matters. In other words, presents are *distinguishable* while pennies are not. (A question like that in section 3.2 where we specify in advance how many presents a given child gets would be trivial for money: There is only one way to distribute n pennies so that the first child gets n_1, the second child gets n_2, etc.)

Even though the problem is quite different from the distribution of presents, we can solve it by imagining a similar distribution method. We line up the pennies (it does not matter in which order; they are all alike), and then let the first child begin to pick them up from left to right. After a while we stop him and let the second child pick up pennies, etc. (Figure 3.2). *The distribution of the money is determined by specifying where to start with a new child.*

Now, there are $n - 1$ points (between consecutive pennies) where we can let a new child in, and we have to select $k - 1$ of them (since the first child always starts at the beginning, we have no choice there). Thus we have to select a $(k - 1)$-element subset from an $(n - 1)$-element set. The number of ways of doing so is $\binom{n-1}{k-1}$.

To sum up, we have the following theorem:

Theorem 3.4.1 *The number of ways to distribute n identical pennies to k children so that each child gets at least one is $\binom{n-1}{k-1}$.*

It is quite surprising that the binomial coefficients give the answer here, in a quite nontrivial and unexpected way!

Let's also discuss the natural (though unfair) modification of this question, where we also allow distributions in which some children get no money at all; we consider even giving all the money to one child. With the following trick, we can reduce the problem of counting such distributions to the problem we just solved: We borrow 1 penny from each child, and then distribute the whole amount (i.e., $n + k$ pennies) to the children so that each child gets at least one penny. This way every child gets back the money we borrowed from him or her, and the lucky ones get some more. The "more" is exactly n pennies distributed to k children. We already know that the number of ways to distribute $n + k$ pennies to k children so that each child gets at least one penny is $\binom{n+k-1}{k-1}$. So we have the next result:

Theorem 3.4.2 *The number of ways to distribute n identical pennies to k children is $\binom{n+k-1}{k-1}$.*

3.4.1 In how many ways can you distribute n pennies to k children if each child is supposed to get at least 2?

3.4.2 We distribute n pennies to k boys and ℓ girls in such a way that (to be really unfair) we require that each of the girls gets at least one penny (but we do not insist on the same thing for the boys). In how many ways can we do this?

3.4.3 A group of k earls are playing cards. Originally, they each have p pennies. At the end of the game, they count how much money they have. They do not borrow from each other, so that each cannot loose more than his p pennies. How many possible results are there?

3.5   Pascal's Triangle

To study various properties of binomial coefficients, the following picture is very useful. We arrange all binomial coefficients into a triangular scheme: in the "zeroth" row we put $\binom{0}{0}$; in the first row, we put $\binom{1}{0}$ and $\binom{1}{1}$; in the second row, $\binom{2}{0}$, $\binom{2}{1}$, and $\binom{2}{2}$; etc. In general, the nth row contains the numbers $\binom{n}{0}, \binom{n}{1}, \ldots, \binom{n}{n}$. We shift these rows so that their midpoints match; this way we get a pyramidlike scheme, called *Pascal's Triangle* (named after the French mathematician and philosopher Blaise Pascal, 1623–1662). The figure below shows only a finite piece of Pascal's Triangle.

$$
\begin{array}{ccccccccccccc}
 & & & & & & \binom{0}{0} & & & & & & \\
 & & & & & \binom{1}{0} & & \binom{1}{1} & & & & & \\
 & & & & \binom{2}{0} & & \binom{2}{1} & & \binom{2}{2} & & & & \\
 & & & \binom{3}{0} & & \binom{3}{1} & & \binom{3}{2} & & \binom{3}{3} & & & \\
 & & \binom{4}{0} & & \binom{4}{1} & & \binom{4}{2} & & \binom{4}{3} & & \binom{4}{4} & & \\
 & \binom{5}{0} & & \binom{5}{1} & & \binom{5}{2} & & \binom{5}{3} & & \binom{5}{4} & & \binom{5}{5} & \\
\binom{6}{0} & & \binom{6}{1} & & \binom{6}{2} & & \binom{6}{3} & & \binom{6}{4} & & \binom{6}{5} & & \binom{6}{6}
\end{array}
$$

We can replace each binomial coefficient by its numerical value to get another version of Pascal's Triangle (going a little further down, to the eighth row):

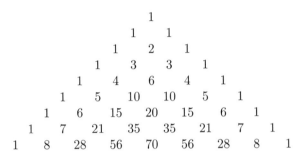

$$
\begin{array}{ccccccccccccccc}
 & & & & & & & 1 & & & & & & & \\
 & & & & & & 1 & & 1 & & & & & & \\
 & & & & & 1 & & 2 & & 1 & & & & & \\
 & & & & 1 & & 3 & & 3 & & 1 & & & & \\
 & & & 1 & & 4 & & 6 & & 4 & & 1 & & & \\
 & & 1 & & 5 & & 10 & & 10 & & 5 & & 1 & & \\
 & 1 & & 6 & & 15 & & 20 & & 15 & & 6 & & 1 & \\
1 & & 7 & & 21 & & 35 & & 35 & & 21 & & 7 & & 1 \\
\end{array}
$$
$$
1 \quad 8 \quad 28 \quad 56 \quad 70 \quad 56 \quad 28 \quad 8 \quad 1
$$

3.5.1 Prove that Pascal's Triangle is symmetric with respect to the vertical line through its apex.

3.5.2 Prove that each row of Pascal's Triangle starts and ends with 1.

3.6   Identities in Pascal's Triangle

Looking at Pascal's Triangle, it is not hard to notice its most important property: Every number in it (other than the 1's on the boundary) is the sum of the two numbers immediately above it. This, in fact, is a property of the binomial coefficients we already met, namely, equation (1.8) in Section 1.8:
$$
\binom{n}{k} = \binom{n-1}{k-1} + \binom{n-1}{k}. \tag{3.2}
$$

This property of Pascal's Triangle enables us to generate the triangle very fast, building it up row by row, using (3.2). It also gives us a tool to prove many properties of the binomial coefficients, as we shall see.

As a first application, let us give a new solution to exercise 3.1.2. There the task was to prove the identity
$$
\binom{n}{0} - \binom{n}{1} + \binom{n}{2} - \binom{n}{3} + \cdots + (-1)^n \binom{n}{n} = 0, \tag{3.3}
$$

using the Binomial Theorem. Now we give a proof based on (3.2): We can replace $\binom{n}{0}$ by $\binom{n-1}{0}$ (both are just 1), $\binom{n}{1}$ by $\binom{n-1}{0} + \binom{n-1}{1}$, $\binom{n}{2}$ by $\binom{n-1}{1} + \binom{n-1}{2}$, etc. Thus we get the sum
$$
\binom{n-1}{0} - \left[\binom{n-1}{0} + \binom{n-1}{1} \right] + \left[\binom{n-1}{1} + \binom{n-1}{2} \right]
$$

$$+ \cdots + (-1)^{n-1} \left[\binom{n-1}{n-2} + \binom{n-1}{n-1} \right] + (-1)^n \binom{n-1}{n-1},$$

which is clearly 0, since the second term in each pair of brackets cancels with the first term in the next pair of brackets.

This method gives more than just a new proof of an identity we already know. What do we get if we start the same way, adding and subtracting binomial coefficients alternatingly, but stop earlier? Writing this as a formula, we take

$$\binom{n}{0} - \binom{n}{1} + \binom{n}{2} - \binom{n}{3} + \cdots + (-1)^k \binom{n}{k}.$$

If we do the same trick as above, we get

$$\binom{n-1}{0} - \left[\binom{n-1}{0} + \binom{n-1}{1} \right] + \left[\binom{n-1}{1} + \binom{n-1}{2} \right] - \cdots$$

$$+ (-1)^k \left[\binom{n-1}{k-1} + \binom{n-1}{k} \right].$$

Here again every term cancels except the last one; so the result is $(-1)^k \binom{n-1}{k}$.

There are many other surprising relations satisfied by the numbers in Pascal's Triangle. For example, let's ask, what is the sum of the *squares* of elements in each row?

Let's experiment by computing the sum of the squares of elements in the first few rows:

$$1^2 = 1,$$
$$1^2 + 1^2 = 2,$$
$$1^2 + 2^2 + 1^2 = 6,$$
$$1^2 + 3^2 + 3^2 + 1^2 = 20,$$
$$1^2 + 4^2 + 6^2 + 4^2 + 1^2 = 70.$$

We may recognize these numbers as the numbers in the middle column of Pascal's Triangle. Of course, only every second row contains an entry in the middle column, so the last value above, the sum of squares in the fourth row is the middle element in the eighth row. So the examples above suggest the following identity:

$$\binom{n}{0}^2 + \binom{n}{1}^2 + \binom{n}{2}^2 + \cdots + \binom{n}{n-1}^2 + \binom{n}{n}^2 = \binom{2n}{n}. \tag{3.4}$$

Of course, the few experiments above do not prove that this identity always holds, so we need a proof.

We will give an interpretation of both sides of the identity as the result of a counting problem; it will turn out that they count the same things, so they are equal. It is obvious what the right-hand side counts: the number of subsets of size n of a set of size $2n$. It will be convenient to choose the set $S = \{1, 2, \ldots, 2n\}$ as our $2n$-element set.

The combinatorial interpretation of the left-hand side is not so easy. Consider a typical term, say $\binom{n}{k}^2$. We claim that this is the number of n-element subsets of $\{1, 2, \ldots, 2n\}$ that contain exactly k elements from $\{1, 2, \ldots, n\}$ (the first half of our set S). In fact, how do we choose such an n-element subset of S? We choose k elements from $\{1, 2, \ldots, n\}$ and then $n - k$ elements from $\{n + 1, n + 2, \ldots, 2n\}$. The first can be done in $\binom{n}{k}$ ways; no matter which k-element subset of $\{1, 2, \ldots, n\}$ we selected, we have $\binom{n}{n-k}$ ways to choose the other part. Thus the number of ways to choose an n-element subset of S having k elements from $\{1, 2, \ldots, n\}$ is

$$\binom{n}{k} \cdot \binom{n}{n-k} = \binom{n}{k}^2$$

(by the symmetry of Pascal's Triangle).

Now, to get the total number of n-element subsets of S, we have to sum these numbers for all values of $k = 0, 1, \ldots, n$. This proves identity (3.4).

3.6.1 Give a proof of the formula (1.9),

$$1 + \binom{n}{1} + \binom{n}{2} + \cdots + \binom{n}{n-1} + \binom{n}{n} = 2^n,$$

along the lines of the proof of (3.3). (One could expect that, as with the "alternating" sum, we could get a nice formula for the sum obtained by stopping earlier, like $\binom{n}{0} + \binom{n}{1} + \cdots + \binom{n}{k}$. But this is not the case: No simpler expression is known for this sum in general.)

3.6.2 By the Binomial Theorem, the right-hand side in identity (3.4) is the coefficient of $x^n y^n$ in the expansion of $(x + y)^{2n}$. Write $(x + y)^{2n}$ in the form $(x + y)^n (x + y)^n$, expand both factors $(x + y)^n$ using the Binomial Theorem, and then try to figure out the coefficient of $x^n y^n$ in the product. Show that this gives another proof of identity (3.4).

3.6.3 Prove the following identity:

$$\binom{n}{0}\binom{m}{k} + \binom{n}{1}\binom{m}{k-1} + \cdots + \binom{n}{k-1}\binom{m}{1} + \binom{n}{k}\binom{m}{0} = \binom{n+m}{k}.$$

You can use a combinatorial interpretation of both sides, as in the proof of (3.4) above, or the Binomial Theorem as in the previous exercise.

Here is another relation between the numbers in Pascal's Triangle. Let us start with the first element in the nth row, and sum the elements moving down diagonally to the right (Figure 3.3). For example, starting with the first element in the second row, we get

$$1 = 1,$$
$$1 + 3 = 4,$$
$$1 + 3 + 6 = 10,$$
$$1 + 3 + 6 + 10 = 20,$$
$$1 + 3 + 6 + 10 + 15 = 35.$$

These numbers are just the numbers in the next skew line of the table!

$$
\begin{array}{ccccccccccccc}
 & & & & & & 1 & & & & & & \\
 & & & & & 1 & & 1 & & & & & \\
 & & & & \mathbf{1} & & 2 & & 1 & & & & \\
 & & & 1 & & \mathbf{3} & & 3 & & 1 & & & \\
 & & 1 & & 4 & & \mathbf{6} & & 4 & & 1 & & \\
 & 1 & & 5 & & 10 & & \mathbf{10} & & 5 & & 1 & \\
 1 & & 6 & & 15 & & 20 & & \mathbf{15} & & 6 & & 1 \\
\end{array}
$$

$$
\begin{array}{ccccccccc}
1 & 7 & 21 & 35 & 35 & \mathbf{21} & 7 & 1 & \\
1 & 8 & 28 & 56 & 70 & \boxed{\mathbf{56}} & 28 & 8 & 1 \\
\end{array}
$$

FIGURE 3.3. Adding up entries in Pascal's Triangle diagonally.

If we want to put this in a formula, we get

$$\binom{n}{0} + \binom{n+1}{1} + \binom{n+2}{2} + \cdots + \binom{n+k}{k} = \binom{n+k+1}{k}. \tag{3.5}$$

To *prove* this identity, we use induction on k. If $k = 0$, the identity just says that $1 = 1$, so it is trivially true. (We can check it also for $k = 1$, even though this is not necessary. Anyway, it says that $1 + (n + 1) = n + 2$.)

So suppose that the identity (3.5) is true for a given value of k, and we want to prove that it also holds for $k + 1$ in place of k. In other words, we want to prove that

$$\binom{n}{0} + \binom{n+1}{1} + \binom{n+2}{2} + \cdots + \binom{n+k}{k} + \binom{n+k+1}{k+1} = \binom{n+k+2}{k+1}.$$

Here the sum of the first k terms on the left-hand side is $\binom{n+k+1}{k}$ by the induction hypothesis, and so the left-hand side is equal to

$$\binom{n+k+1}{k} + \binom{n+k+1}{k+1}.$$

But this is indeed equal to $\binom{n+k+2}{k+1}$ by the fundamental property (3.2) of Pascal's Triangle. This completes the proof by induction.

3.6.4 Suppose that you want to choose a $(k+1)$-element subset of the $(n+k+1)$-element set $\{1, 2, \ldots, n+k+1\}$. You decide to do this by choosing first the largest element, then the rest. Show that counting the number of ways to choose the subset this way, you get a combinatorial proof of identity (3.5).

3.7   A Bird's-Eye View of Pascal's Triangle

Let's imagine that we are looking at Pascal's Triangle from a distance. Or to put it differently, we are not interested in the exact numerical values of the entries, but rather in their order of magnitude, rise and fall, and other global properties. The first such property of Pascal's Triangle is its symmetry (with respect to the vertical line through its apex), which we already know.

Another property one observes is that *along any row, the entries increase until the middle, and then decrease*. If n is even, there is a unique middle element in the nth row, and this is the largest; if n is odd, then there are two equal middle elements, which are largest.

So let us *prove* that the entries increase until the middle (then they begin to decrease by the symmetry of the table). We want to compare two consecutive entries:

$$\binom{n}{k} \; ? \; \binom{n}{k+1}.$$

If we use the formula in Theorem 1.8.1, we can write this as

$$\frac{n(n-1)\cdots(n-k+1)}{k(k-1)\cdots 1} \; ? \; \frac{n(n-1)\cdots(n-k)}{(k+1)k\cdots 1}.$$

There are many common factors on both sides that are positive, and so we can simplify. We get the really simple comparison

$$1 \; ? \; \frac{n-k}{k+1}.$$

Rearranging, we get

$$k \; ? \; \frac{n-1}{2}.$$

So if $k < (n-1)/2$, then $\binom{n}{k} < \binom{n}{k+1}$; if $k = (n-1)/2$, then $\binom{n}{k} = \binom{n}{k+1}$ (this is the case of the two entries in the middle if n is odd); and if $k > (n-1)/2$, then $\binom{n}{k} > \binom{n}{k+1}$.

It will be useful later that this computation also describes by *how much* consecutive elements increase or decrease. If we start from the left, the second entry (namely, n) is larger by a factor of n than the first; the third (namely, $n(n-1)/2$) is larger by a factor of $(n-1)/2$ than the second. In general,

$$\frac{\binom{n}{k+1}}{\binom{n}{k}} = \frac{n-k}{k+1}. \tag{3.6}$$

3.7.1 For which values of n and k is $\binom{n}{k+1}$ twice the previous entry in Pascal's Triangle?

3.7.2 Instead of the ratio, look at the difference of two consecutive entries in Pascal's Triangle:

$$\binom{n}{k+1} - \binom{n}{k}.$$

For which value of k is this difference largest?

We know that each row of Pascal's Triangle is symmetric. We also know that the entries start with 1, rise to the middle, and then fall back to 1. Can we say more about their shape?

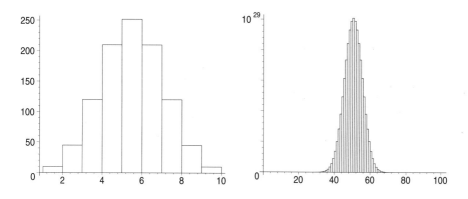

FIGURE 3.4. Bar chart of the nth row of Pascal's Triangle, for $n = 10$ and $n = 100$.

Figure 3.4 shows the graph of the numbers $\binom{n}{k}$ $(k = 0, 1, \ldots, n)$ for the values $n = 10$ and $n = 100$. We can make several further observations.

— First, the largest number gets very large.

— Second, not only do these numbers increase to the middle and then decrease, but the middle ones are substantially larger than those at the beginning and end. For $n = 100$, we see bars only in the range $\binom{100}{25}, \binom{100}{26}, \ldots, \binom{100}{75}$; the numbers outside this range are so small compared to the largest that they do not show in the figure.

— Third, we can observe that the shape of the graph is quite similar for different values of n.

Let's look more carefully at these observations. For the discussions that follow, we shall assume that n is even (for odd values of n, the results would be quite similar, except that one would have to word them differently). If

n is even, then we already know that the largest entry in the nth row is the middle number $\binom{n}{n/2}$, and all other entries are smaller.

How large is the largest number in the nth row of Pascal's Triangle? We know immediately an upper bound on this number:

$$\binom{n}{n/2} < 2^n,$$

since 2^n is the sum of all entries in the row. It only takes a little more sophistication to get a lower bound:

$$\binom{n}{n/2} > \frac{2^n}{n+1},$$

since $2^n/(n+1)$ is the average of the numbers in the row, and the largest number is certainly at least as large as the average.

These bounds already give a pretty good idea about the size of $\binom{n}{n/2}$; in particular, they show that this number gets very large. Take, say, $n = 500$. Then we get

$$\frac{2^{500}}{501} < \binom{500}{250} < 2^{500}.$$

If we want to know the number of digits of $\binom{500}{250}$, we just have to take the logarithm (in base 10) of it. From the bounds above, we get

$$500 \lg 2 - \lg 501 = 147.8151601 \cdots < \lg \binom{500}{250} < 500 \lg 2 = 150.5149978 \ldots .$$

This inequality gives the number of digits with a small error: If we guess that it is 150, then we are off by at most 2 (actually, 150 is the true value).

Using Stirling's formula (Theorem 2.2.1), one can get an even better approximation of this largest entry. We know that

$$\binom{n}{n/2} = \frac{n!}{(n/2)!(n/2)!}.$$

Here, by the Stirling's formula,

$$n! \sim \sqrt{2\pi n}\left(\frac{n}{e}\right)^n, \qquad (n/2)! \sim \sqrt{\pi n}\left(\frac{n}{2e}\right)^{n/2},$$

and so

$$\binom{n}{n/2} \sim \frac{\sqrt{2\pi n}\left(\frac{n}{e}\right)^n}{\pi n\left(\frac{n}{2e}\right)^n} = \sqrt{\frac{2}{\pi n}}2^n. \tag{3.7}$$

So we know that the largest entry in the nth row of Pascal's Triangle is in the middle, and we know approximately how large this element is. We also know that going either left or right, the elements begin to drop. How fast do they drop? Figure 3.4 suggests that starting from the middle, the binomial coefficients drop just by a little at the beginning, but pretty soon this accelerates.

Looking from the other end, we see this even more clearly. Let us consider, say, row 57 (just to take a non-round number for a change). The first few elements are

1, 57, 1596, 29260, 395010, 4187106, 36288252, 264385836, 1652411475,

8996462475, 43183019880, 184509266760, 707285522580, ...

and the ratios between consecutive entries are:

57, 28, 18.33, 13.5, 10.6, 8.67, 7.29, 6.25, 5.44, 4.8, 4.27, 3.83, ...

While the entries are growing fast, these ratios get smaller and smaller, and we know that when we reach the middle, they have to turn less than 1 (since the entries themselves begin to decrease). But what are these ratios? We computed them above, and found that

$$\frac{\binom{n}{k+1}}{\binom{n}{k}} = \frac{n-k}{k+1}.$$

If we write this as

$$\frac{n-k}{k+1} = \frac{n+1}{k+1} - 1,$$

then we see immediately that the ratio of two consecutive binomial coefficients decreases as k increases.

3.8   An Eagle's-Eye View: Fine Details

Let us ask a more quantitative question about the shape of a row in Pascal's Triangle: Which binomial coefficient in this row is (for example) half of the largest?

We consider the case where n is even; then we can write $n = 2m$, where m is a positive integer. The largest, middle entry in the nth row is $\binom{2m}{m}$. Consider the binomial coefficient that is t steps from the middle. It does not matter whether we go left or right, so take, say, $\binom{2m}{m-t}$. We want to compare it with the largest coefficient.

The following formula describes the rate at which the binomial coefficients drop:

$$\binom{2m}{m-t} \bigg/ \binom{2m}{m} \approx e^{-t^2/m}. \tag{3.8}$$

The graph of right-hand side of (3.8) (as a function of t) is shown in Figure 3.5 for $m = 50$. This is the famous *Gauss curve* (sometimes also called the

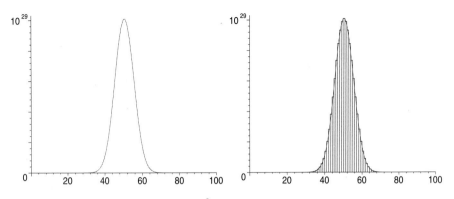

FIGURE 3.5. The Gauss curve $e^{-t^2/m}$ for $m = 50$, and the chart of binomial coefficients in the 100th row of Pascal's Triangle.

"bell curve"). Plotting many types of statistics gives a similar picture. In Figure 3.5 we show the curve alone and also overlaid with the binomial coefficients, to show the excellent fit.

Equation (3.8) is not an exact equation, and to make it a precise mathematical statement, we need to tell how large the error can be. Below, we shall derive the following inequalities:

$$e^{-t^2/(m-t+1)} \leq \binom{2m}{m-t} \bigg/ \binom{2m}{m} \leq e^{-t^2/(m+t)} . \tag{3.9}$$

The upper and lower bounds in this formula are quite similar to the (imprecise) approximation $e^{-t^2/m}$ given in (3.8), and it is easy to see that the latter value is between them. The right hand side of (3.8) in fact gives a better approximation than either the upper or the lower bound. For example, suppose that we want to estimate the ratio $\binom{100}{40} / \binom{100}{50}$, which is $0.1362\ldots$. From (3.9) we get

$$0.08724 \leq \binom{100}{40} \bigg/ \binom{100}{50} \leq 0.1889,$$

while the approximation given in (3.8) is $0.1353\ldots$, much closer to the truth. Using heavier calculus (analysis) would give tighter bounds; we give here only as much as we can without appealing to calculus.

To derive (3.9), we start with transforming the ratio in the middle; or rather, we take its reciprocal, which is larger than 1 and therefore a bit easier to work with:

$$\binom{2m}{m} \bigg/ \binom{2m}{m-t} = \frac{(2m)!}{m!m!} \bigg/ \frac{(2m)!}{(m-t)!(m+t)!} = \frac{(m-t)!(m+t)!}{m!m!}$$
$$= \frac{(m+t)(m+t-1)\cdots(m+1)}{m(m-1)\cdots(m-t+1)}.$$

So we have some sort of a formula for this ratio, but how useful is it? How do we tell, for example, for which value of t this ratio becomes larger than 2? We can certainly write this as a formula:

$$\frac{(m+t)(m+t-1)\cdots(m+1)}{m(m-1)\cdots(m-t+1)} > 2. \tag{3.10}$$

We could try to solve this inequality for t (similar to the proof that the entries are increasing to the middle), but it would be too complicated to solve. So even to answer such a simple question about binomial coefficients like, how far from the middle do they drop to half of the maximum? needs more work, and we have to do some arithmetic trickery. We divide the first factor of the numerator by the first factor of the denominator, the second factor by the second factor etc., to get

$$\frac{m+t}{m} \cdot \frac{m+t-1}{m-1} \cdots \frac{m+1}{m-t+1}.$$

This product is still not easy to handle, but we have met similar ones in Section 2.5! There the trick was to take the logarithm, and this works here just as well. We get

$$\ln\left(\frac{m+t}{m}\right) + \ln\left(\frac{m+t-1}{m-1}\right) + \cdots + \ln\left(\frac{m+1}{m-t+1}\right).$$

Just as in Section 2.5, we can estimate the logarithms on the left-hand side using the inequalities in Lemma 2.5.1. Let's start with deriving an upper bound. For a typical term in the sum we have

$$\ln\left(\frac{m+t-k}{m-k}\right) \le \frac{m+t-k}{m-k} - 1 = \frac{t}{m-k},$$

and so

$$\ln\left(\frac{m+t}{m}\right) + \ln\left(\frac{m+t-1}{m-1}\right) + \cdots + \ln\left(\frac{m+1}{m-t+1}\right)$$
$$\le \frac{t}{m} + \frac{t}{m-1} + \cdots + \frac{t}{m-t+1}.$$

Can we bring this sum into a closed form? No, but we can use another trick from section 2.5. We replace each denominator by $m-t+1$, since this is the smallest; then we increase all the fractions (except the last one, which we don't change) and get an upper bound:

$$\frac{t}{m} + \frac{t}{m-1} + \cdots + \frac{t}{m-t+1} \le \frac{t}{m-t+1} + \frac{t}{m-t+1} + \cdots + \frac{t}{m-t+1}$$
$$= \frac{t^2}{m-t+1}.$$

Remember, this is an upper bound on the *logarithm* of the ratio $\binom{2n}{m} / \binom{2m}{m-t}$; to get an upper bound on the ratio itself, we have to apply the exponential function. Then we have another step to undo: we have to take the reciprocal, to get the lower bound in (3.9).

The upper bound in (3.9) can be derived using similar steps; the details are left to the reader as an exercise 3.8.2.

Let us return to our earlier question: We want to know when (for which value of t) the quotient in (3.9) will be larger than 2. We might need similar information for other numbers instead of 2, so let's try to answer the question for a general number $C > 1$. Thus we want to know for which value of t we get

$$\binom{2m}{m} / \binom{2m}{m-t} > C. \tag{3.11}$$

From (3.8) we know that the left-hand side is about $e^{t^2/m}$, so we start with solving the equation

$$e^{t^2/m} = C.$$

The exponential function on the left looks nasty, but the good old logarithm helps again: We get

$$\frac{t^2}{m} = \ln C,$$

which is now easy to solve:

$$t = \sqrt{m \ln C}.$$

So we expect that if t is larger than this, than (3.11) holds. But of course we have to be aware of the fact that this is only an approximation, not a precise result! Instead of (3.8), we can use the exact inequalities (3.9) to get the following lemma:

Lemma 3.8.1 *If $t \geq \sqrt{m \ln C} + \ln C$, then (3.11) holds; if $t \leq \sqrt{m \ln C} - \ln C$, then (3.11) does not hold.*

The derivation of these conditions from (3.9) is similar to the derivation of the approximate result from (3.8) and is left to the reader as exercise 3.8.3 (difficult!).

In important applications of binomial coefficients (one of which, the law of large numbers, will be discussed in Chapter 5) we also need to know a good bound on the sum of the smallest binomial coefficients, compared with the sum of all of them. Luckily, our previous observations and lemmas enable us to get an answer with some computation but without substantial new ideas.

Lemma 3.8.2 *Let $0 \le k \le m$ and $c = \binom{2m}{k} / \binom{2m}{m}$. Then*

$$\binom{2m}{0} + \binom{2m}{1} + \cdots + \binom{2m}{k-1} < \frac{c}{2} \cdot 2^{2m}. \tag{3.12}$$

To digest the meaning of this, choose $m = 500$, and let's try to see how many binomial coefficients in the 1000th row we have to add up (starting with $\binom{1000}{0}$) to reach 0.5% of the total. Lemma 3.8.2 tells us that if we choose $0 \le k \le 500$ so that $\binom{1000}{k} / \binom{1000}{500} < 1/100$, then adding up the first k binomial coefficients gives a sum less than 0.5% of the total. In turn, Lemma 3.8.1 tells us a k that will be certainly good: any $k \le 500 - \sqrt{500 \ln 100} - \ln 100 = 447.4$. So the first 447 entries in the 1000th row of Pascal's Triangle make up less than 0.5% of the total sum. By the symmetry of Pascal's Triangle, the last 447 add up to another less than 0.5%. The middle 107 terms account for 99% of the total.

Proof. To prove this lemma, let us write $k = m - t$, and compare the sum on the left-hand side of (3.12) with the sum

$$\binom{2m}{m-t} + \binom{2m}{m-t+1} + \cdots + \binom{2m}{m-1}. \tag{3.13}$$

Let us denote the sum $\binom{2m}{0} + \binom{2m}{1} + \cdots + \binom{2m}{m-t-1}$ by A, and the sum $\binom{2m}{m-t} + \binom{2m}{m-t+1} + \cdots + \binom{2m}{m-1}$ by B.

We have

$$\binom{2m}{m-t} = c\binom{2m}{m}$$

by the definition of c. This implies that

$$\binom{2m}{m-t-1} < c\binom{2m}{m-1},$$

since we know that binomial coefficients drop by a larger factor from $\binom{2m}{m-t}$ to $\binom{2m}{m-t-1}$ than they do from $\binom{2m}{m}$ to $\binom{2m}{m-1}$. Repeating the same argument,[1] we get that

$$\binom{2m}{m-t-i} < c\binom{2m}{m-i}$$

for every $i \ge 0$.

Hence it follows that the sum of any t consecutive binomial coefficients is less than c times the sum of the next t (as long as these are all on the left hand side of Pascal's Triangle). Going back from $\binom{2m}{m-1}$, the first block of t binomial coefficients adds up to A (by the definition of A); the next block

---

[1] In other words, using induction.

of t adds up to less than cA, the next block to less than $c^2 A$, etc. Adding up, we get that

$$B < cA + c^2 A + c^3 A \ldots$$

On the right-hand side we only have to sum $\lceil (m - t)/t \rceil$ terms, but we are generous and let the summation run to infinity! The geometric series on the right-hand side adds up to $\frac{c}{1-c} A$, so we get that

$$B < \frac{c}{1 - c} A.$$

We need another inequality involving A and B, but this is easy:

$$B + A < \frac{1}{2} 2^{2m}$$

(since the sum on the left-hand side includes only the left-hand side of Pascal's Triangle, and the middle element is not even counted). From these two inequalities we get

$$B < \frac{c}{1 - c} A < \frac{c}{1 - c} \left(\frac{1}{2} 2^{2m} - B \right),$$

and hence

$$\left(1 + \frac{c}{1 - c} \right) B < \frac{c}{1 - c} \frac{1}{2} 2^{2m}.$$

Multiplying by $1 - c$ gives that $B < c\frac{1}{2} 2^{2m}$, which proves the lemma.   □

3.8.1   (a) Check that the approximation in (3.8) is always between the lower and upper bounds given in (3.9).

(b) Let $2m = 100$ and $t = 10$. By what percentage is the upper bound in (3.9) larger than the lower bound?

3.8.2 Prove the upper bound in (3.9).

3.8.3 Complete the proof of Lemma 3.8.1.

Review Exercises

3.8.4 Find all values of n and k for which $\binom{n}{k+1} = 3\binom{n}{k}$.

3.8.5 Find the value of k for which $k\binom{99}{k}$ is largest.

3.8.6 In city with a regular "chessboard" street plan, the North-South streets are called 1st Street, 2nd Street, ..., 20th Street, and the East-West streets are called 1st Avenue, 2nd Avenue, ..., 10th Avenue. What is the minimum number of blocks you have to walk to get from the corner of 1st Street and 1st Avenue to the corner of 20th Street and 10th Avenue? In how many ways can you get there walking this minimum number of blocks?

3.8.7 In how many ways can you read off the word MATHEMATICS from the following tables:

M	A	T	H	E	M		M	A	T	H		
A	T	H	E	M	A		A	T	H	E		
T	H	E	M	A	T		T	H	E	M	A	
H	E	M	A	T	I		H	E		A	T	I
E	M	A	T	I	C		M	A		T	I	C
M	A	T	I	C	S				I	C	S	

3.8.8 Prove the following identities:

$$\sum_{k=0}^{m}(-1)^k\binom{n}{k} = (-1)^m\binom{n-1}{m};$$

$$\sum_{k=0}^{n}\binom{n}{k}\binom{k}{m} = \binom{n}{m}2^{n-m}.$$

3.8.9 Prove the following inequalities:

$$\frac{n^k}{k^k} \leq \binom{n}{k} \leq \frac{n^k}{k!}.$$

3.8.10 In how many ways can you distribute n pennies to k children if each child is supposed to get at least 5?

3.8.11 Prove that if we move straight down in Pascal's Triangle (visiting every other row), then the numbers we see are increasing.

3.8.12 Prove that

$$1 + \binom{n}{1}2 + \binom{n}{2}4 + \cdots + \binom{n}{n-1}2^{n-1} + \binom{n}{n}2^n = 3^n.$$

Try to find a combinatorial proof.

3.8.13 Suppose that you want to choose a $(2k+1)$-element subset of the n-element set $\{1, 2, \ldots, n\}$. You decide to do this by choosing first the middle element, then the k elements to its left, then the k elements to its right. Formulate the combinatorial identity you get from this.

3.8.14 Let n be a positive integer divisible by 3. Use Stirling's formula to find the approximate value of $\binom{n}{n/3}$.

3.8.15 Prove that $\binom{n}{10} \sim \frac{n^{10}}{10!}$.

4
Fibonacci Numbers

4.1 Fibonacci's Exercise

In the thirteenth century, the Italian mathematician Leonardo Fibonacci studied the following (not too realistic) question:

Leonardo Fibonacci

A farmer raises rabbits. Each rabbit gives birth to one rabbit when it turns 2 months old, and then to one rabbit each month thereafter. Rabbits never die, and we ignore male rabbits. How many rabbits will the farmer have in the nth month if he starts with one newborn rabbit?

It is easy to figure out the answer for small values of n. The farmer has 1 rabbit in the first month and 1 rabbit in the second month, since the rabbit has to be 2 months old before starting to reproduce. He has 2 rabbits during the third month, and 3 rabbits during the fourth, since his first rabbit delivered a new one after the second and one after the third. After 4 months, the second rabbit also delivers a new rabbit, so two new rabbits are added. This means that the farmer will have 5 rabbits during the fifth month.

It is easy to follow the multiplication of rabbits for any number of months if we notice that the number of new rabbits added each months is just the same as the number of rabbits who are at least 2 months old, i.e., who were already there in the previous month. In other words, to get the number of rabbits in the *next* month, we have to add the number of rabbits in the *previous* month to the number of rabbits in the *current* month. This makes it easy to compute the numbers one by one:

$$1, \ 1, \ 1 + 1 = 2, \ 2 + 1 = 3, \ 3 + 2 = 5, \ 5 + 3 = 8, \ 8 + 5 = 13, \ \dots$$

(It is quite likely that Fibonacci did not get his question as a real applied math problem; he played with numbers, noticed that this procedure gives numbers that were new to him but nevertheless had very interesting properties—as we'll see ourselves—and then tried to think of an "application.")

To write this as a formula, let us denote by F_n the number of rabbits during the nth month. Then we have, for $n = 2, 3, 4, \dots$,

$$F_{n+1} = F_n + F_{n-1}. \tag{4.1}$$

We also know that $F_1 = 1$, $F_2 = 1$, $F_3 = 2$, $F_4 = 3$, $F_5 = 5$. It is convenient to define $F_0 = 0$; then equation (4.1) will remain valid for $n = 1$ as well. Using equation (4.1), we can easily determine any number of terms in this sequence of numbers:

$$0, \ 1, \ 1, \ 2, \ 3, \ 5, \ 8, \ 13, \ 21, \ 34, \ 55, \ 89, \ 144, \ 233, \ 377, \ 610, \ 987, \ 1597, \dots .$$

The numbers in this sequence are called *Fibonacci numbers*.

We see that equation (4.1), together with the special values $F_0 = 0$ and $F_1 = 1$, uniquely determines the Fibonacci numbers. Thus we can consider (4.1), together with $F_0 = 0$ and $F_1 = 1$, as the definition of these numbers. This may seem a somewhat unusual definition: Instead of telling what F_n is (say, by a formula), we just give a rule that computes each Fibonacci number from the two previous numbers, and specify the first two values. Such a definition is called a *recurrence*. It is quite similar to induction in spirit (except that it is not a proof technique, but a definition method), and is sometimes also called *definition by induction*.

4.1.1 Why do we have to specify exactly two of the elements to begin with? Why not one or three?

Before trying to say more about these numbers, let us consider another counting problem:

A staircase has n steps. You walk up taking one or two at a time. How many ways can you go up?

For $n = 1$, there is only 1 way. For $n = 2$, you have 2 choices: take one step twice or two once. For $n = 3$, you have 3 choices: three single steps, or one single followed by one double, or one double followed by one single.

Now stop and try to guess what the answer is in general! If you guessed that the number of ways to go up on a stair with n steps is n, you are wrong. The next case, $n = 4$, gives 5 possibilities $(1+1+1+1, 2+1+1, 1+2+1, 1+1+2, 2+2)$.

So instead of guessing, let's try the following strategy. Let's denote by J_n the answer. We try to figure out what J_{n+1} is, assuming we know the value of J_k for $1 \le k \le n$. If we start with a single step, we have J_n ways to go up the remaining n steps. If we start with a double step, we have J_{n-1} ways to go up the remaining $n-1$ steps. These are all the possibilities, and so

$$J_{n+1} = J_n + J_{n-1}.$$

This equation is the same as the equation we have used to compute the Fibonacci numbers F_n. Does this means that $F_n = J_n$? Of course not, as we see by looking at the beginning values: for example, $F_3 = 2$ but $J_3 = 3$. However, it is easy to observe that all that happens is that the J_n are shifted by one:

$$J_n = F_{n+1}.$$

This is valid for $n = 1, 2$, and then of course it is valid for every n, since the sequences $F_2, F_3, F_4, \ldots$ and $J_1, J_2, J_3, \ldots$ are computed by the same rule from their first two elements.

4.1.2 We have n dollars to spend. Every day we buy either a candy for 1 dollar or an ice cream for 2 dollars. In how many ways can we spend the money?

4.1.3 How many subsets does the set $\{1, 2, \ldots, n\}$ have that contain no two consecutive integers?

4.2 Lots of Identities

There are many interesting relations valid for the Fibonacci numbers. For example, what is the sum of the first n Fibonacci numbers? We have

$$0 = 0,$$
$$0 + 1 = 1,$$
$$0 + 1 + 1 = 2,$$
$$0 + 1 + 1 + 2 = 4,$$
$$0 + 1 + 1 + 2 + 3 = 7,$$
$$0 + 1 + 1 + 2 + 3 + 5 = 12,$$
$$0 + 1 + 1 + 2 + 3 + 5 + 8 = 20,$$
$$0 + 1 + 1 + 2 + 3 + 5 + 8 + 13 = 33.$$

Staring at these numbers for a while, it is not hard to recognize that by adding 1 to the right-hand sides we get Fibonacci numbers; in fact, we get Fibonacci numbers two steps after the last summand. As a formula, we have

$$F_0 + F_1 + F_2 + \cdots + F_n = F_{n+2} - 1.$$

Of course, at this point this is only a *conjecture*, an unproven mathematical statement we believe to be true. To prove it, we use induction on n (since the Fibonacci numbers are defined by recurrence, induction is the natural and often only proof method at hand).

We have already checked the validity of the statement for $n = 0$ and 1. Suppose that we know that the identity holds for the sum of the first $n - 1$ Fibonacci numbers. Consider the sum of the first n Fibonacci numbers:

$$F_0 + F_1 + \cdots + F_n = (F_0 + F_1 + \cdots + F_{n-1}) + F_n = (F_{n+1} - 1) + F_n,$$

by the induction hypothesis. But now we can use the recurrence equation for the Fibonacci numbers to get

$$(F_{n+1} - 1) + F_n = F_{n+2} - 1.$$

This completes the induction proof.

4.2.1 Prove that F_{3n} is even.

4.2.2 Prove that F_{5n} is divisible by 5.

4.2.3 Prove the following identities.
  (a) $F_1 + F_3 + F_5 + \cdots + F_{2n-1} = F_{2n}.$
  (b) $F_0 - F_1 + F_2 - F_3 + \cdots - F_{2n-1} + F_{2n} = F_{2n-1} - 1.$

(c) $F_0^2 + F_1^2 + F_2^2 + \cdots + F_n^2 = F_n \cdot F_{n+1}$.

(d) $F_{n-1}F_{n+1} - F_n^2 = (-1)^n$.

4.2.4 We want to extend the Fibonacci numbers in the other direction; i.e., we want to define F_n for negative values of n. We want to do this so that the basic recurrence (4.1) remain valid. So from $F_{-1} + F_0 = F_1$ we get $F_{-1} = 1$; then from $F_{-2} + F_{-1} = F_0$ we get $F_{-2} = -1$, etc. How are these "Fibonacci numbers with negative indices" related to those with positive indices? Find several values, conjecture, and then prove the answer.

Now we state a little more difficult identity:

$$F_n^2 + F_{n-1}^2 = F_{2n-1}. \tag{4.2}$$

It is easy to check this for many values of n, and we can be convinced that it is true, but to prove it is a bit more difficult. Why is this more difficult than previous identities? Because if we want to prove it by induction (we don't really have other means at this point), then on the right-hand side we have only every other Fibonacci number, and so we don't know how to apply the recursion there.

One way to fix this is to find a similar formula for F_{2n}, and prove both of them by induction. With some luck (or deep intuition?) you can conjecture the following:

$$F_{n+1}F_n + F_n F_{n-1} = F_{2n}. \tag{4.3}$$

Again, it is easy to check that this holds for many small values of n. To prove (4.3), let us use the basic recurrence (4.1) twice:

$$F_{n+1}F_n + F_n F_{n-1} = (F_n + F_{n-1})F_n + (F_{n-1} + F_{n-2})F_{n-1}$$
$$= \left(F_n^2 + F_{n-1}^2\right) + (F_n F_{n-1} + F_{n-1}F_{n-2})$$

(apply (4.2) to the first term and induction to the second term)

$$= F_{2n-1} + F_{2n-2} = F_{2n}.$$

The proof of (4.2) is similar:

$$F_n^2 + F_{n-1}^2 = (F_{n-1} + F_{n-2})^2 + F_{n-1}^2$$
$$= (F_{n-1}^2 + F_{n-2}^2) + 2F_{n-1}F_{n-2} + F_{n-1}^2$$
$$= (F_{n-1}^2 + F_{n-2}^2) + F_{n-1}(F_{n-2} + F_{n-1}) + F_{n-1}F_{n-2}$$
$$= (F_{n-1}^2 + F_{n-2}^2) + F_n F_{n-1} + F_{n-1}F_{n-2}$$

(apply induction to the first term and (4.3) to the second term)

$$= F_{2n-3} + F_{2n-2} = F_{2n-1}.$$

Wait a minute! What kind of trickery is this? We use (4.3) in the proof of (4.2), and then (4.2) in the proof of (4.3)? Relax, the argument is OK: It is just that the two induction proofs have to go simultaneously. If we know that both (4.3) and (4.2) are true for a certain value of n, then we prove (4.2) for the next value (if you look at the proof, you can see that it uses smaller values of n only), and then use this and the induction hypothesis again to prove (4.3).

This trick is called *simultaneous induction*, and it is a useful method to make induction more powerful.

4.2.5 Prove that the following recurrence can be used to compute Fibonacci numbers of odd index, without computing those with even index:

$$F_{2n+1} = 3F_{2n-1} - F_{2n-3}.$$

Use this identity to prove (4.2) without the trick of simultaneous induction. Give a similar proof of (4.3).

4.2.6 Mark the first entry of any row of Pascal's triangle (this is a 1). Move one step east and one step northeast, and mark the entry there. Repeat this until you get out of the triangle. Compute the sum of the entries you marked.

(a) What numbers do you get if you start from different rows? First conjecture, than prove your answer.

(b) Formulate this fact as an identity involving binomial coefficients.

Suppose that Fibonacci's farmer starts with A newborn rabbits. At the end of the first month (when there is no natural population increase yet), he buys $B - A$ newborn rabbits so that he has B rabbits. From here on, rabbits begin to multiply, and so he has $A + B$ rabbits after the second month, $A + 2B$ rabbits after the third month, etc. How many rabbits will he have after the nth month? Mathematically, we define a sequence $E_0, E_1, E_2, \ldots$ by $E_0 = A$, $E_1 = B$, and from then on, $E_{n+1} = E_n + E_{n-1}$ (the rabbits multiply by the same rule of biology; just the starting numbers are different).

For every two numbers A and B, we have this "modified Fibonacci sequence." How different are they from the real Fibonacci sequence? Do we have to study them separately for every choice of A and B?

It turns out that the numbers E_n can be expressed quite easily in terms of the Fibonacci numbers F_n. To see this, let us compute a few beginning values of the sequence E_n (of course, the result will contain the starting values A and B as parameters).

$$E_0 = A, \quad E_1 = B, \quad E_2 = A + B, \quad E_3 = B + (A + B) = A + 2B,$$
$$E_4 = (A + B) + (A + 2B) = 2A + 3B,$$
$$E_5 = (A + 2B) + (2A + 3B) = 3A + 5B,$$
$$E_6 = (2A + 3B) + (3A + 5B) = 5A + 8B,$$
$$E_7 = (3A + 5B) + (5A + 8B) = 8A + 13B, \ldots .$$

It is easy to recognize what is going on: Each E_n is the sum of a multiple of A and a multiple of B, and the coefficients are ordinary Fibonacci numbers! For a formula, we can conjecture

$$E_n = F_{n-1}A + F_nB. \tag{4.4}$$

Of course, we have not proved this formula; but once we write it up, its proof is so easy (by induction on n, of course) that it is left to the reader as Exercise 4.3.10.

There is an important special case of this identity: We can start with two consecutive Fibonacci numbers $A = F_a$ and $B = F_{a+1}$. Then the sequence E_n is just the Fibonacci sequence, but shifted to the left. Hence we get the following identity:

$$F_{a+b+1} = F_{a+1}F_{b+1} + F_aF_b. \tag{4.5}$$

This is a powerful identity for the Fibonacci numbers, which can be used to derive many others; some applications follow as exercises.

4.2.7 Give a proof of (4.2) and (4.3), based on (4.5).

4.2.8 Use (4.5) to prove the following generalization of Exercises 4.2.1 and 4.2.2: If n is a multiple of k, then F_n is a multiple of F_k.

4.2.9 Cut a chessboard into 4 pieces as shown in Figure 4.1 and assemble a 5×13 rectangle from them. Does this prove that $5 \cdot 13 = 8^2$? Where are we cheating? What does this have to do with Fibonacci numbers?

4.3   A Formula for the Fibonacci Numbers

How large are the Fibonacci numbers? Is there a simple formula that expresses F_n as a function of n?

An easy way out, at least for the author of a book, is to state the answer right away:

Theorem 4.3.1 *The Fibonacci numbers are given by the formula*

$$F_n = \frac{1}{\sqrt{5}} \left(\left(\frac{1 + \sqrt{5}}{2} \right)^n - \left(\frac{1 - \sqrt{5}}{2} \right)^n \right).$$

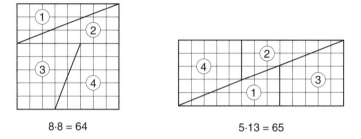

FIGURE 4.1. Proof of $64 = 65$.

Proof. It is straightforward to check that this formula gives the right value for $n = 0, 1$, and then one can prove its validity for all n by induction. $\square$

4.3.1 Prove Theorem 4.3.1 by induction on n.

Do you feel cheated by this proof? You should; while it is, of course, logically correct what we did, one would like to see more: How can one arrive at such a formula? What should we try to get a similar formula if we face a similar, but different, recurrence?

So let us forget Theorem 4.3.1 for a while and let us try to find a formula for F_n "from scratch."

One thing we can try is to experiment. The Fibonacci numbers grow quite fast; how fast? Let's grab our calculator and compute the ratio of consecutive Fibonacci numbers:

$$\frac{1}{1} = 1, \quad \frac{2}{1} = 2, \quad \frac{3}{2} = 1.5, \quad \frac{5}{3} = 1.666666667,$$

$$\frac{8}{5} = 1.600000000, \quad \frac{13}{8} = 1.625000000, \quad \frac{21}{13} = 1.615384615,$$

$$\frac{34}{21} = 1.619047619, \quad \frac{55}{34} = 1.617647059, \quad \frac{89}{55} = 1.618181818,$$

$$\frac{144}{89} = 1.617977528, \quad \frac{233}{144} = 1.618055556, \quad \frac{377}{233} = 1.618025751.$$

It seems that the ratio of consecutive Fibonacci numbers is very close to 1.618, at least if we ignore the first few values. This suggests that the Fibonacci numbers behave like a geometric progression (for a geometric progression, the ratio of any two consecutive elements would be exactly the same). So let's see whether there is any geometric progression that satisfies the same recurrence as the Fibonacci numbers. Let $G_n = c \cdot q^n$ be a geometric progression ($c, q \neq 0$). Then

$$G_{n+1} = G_n + G_{n-1}$$

translates into
$$c \cdot q^{n+1} = c \cdot q^n + c \cdot q^{n-1},$$
which after simplification becomes
$$q^2 = q + 1.$$
So both numbers c and n disappear.[1]

So we have a quadratic equation for q, which we can solve and get
$$q_1 = \frac{1+\sqrt{5}}{2} \approx 1.618034, \qquad q_2 = \frac{1-\sqrt{5}}{2} \approx -0.618034.$$

This gives us two kinds of geometric progressions that satisfy the same recurrence as the Fibonacci numbers:
$$G_n = c\left(\frac{1+\sqrt{5}}{2}\right)^n, \qquad G_n' = c\left(\frac{1-\sqrt{5}}{2}\right)^n$$

(where c is an arbitrary constant). Unfortunately, neither G_n nor G_n' gives the Fibonacci sequence: for one, $G_0 = G_0' = c$, while $F_0 = 0$. But notice that the sequence $G_n - G_n'$ also satisfies the recurrence:
$$G_{n+1} - G_{n+1}' = (G_n + G_{n-1}) - (G_n' + G_{n-1}') = (G_n - G_n') + (G_{n-1} - G_{n-1}')$$

(using that G_n and G_n' satisfy the recurrence). So we have matched the first value F_0, since $G_0 - G_0' = 0$. What about the next one? We have $G_1 - G_1' = c\sqrt{5}$. We can match this with $F_1 = 1$ if we choose $c = 1/\sqrt{5}$.

Thus we have two sequences, F_n and $G_n - G_n'$, that both begin with the same two numbers and satisfy the same recurrence. So we can use the same rule to compute the numbers F_n as the numbers $G_n - G_n'$, and it follows that they must be the same: $F_n = G_n - G_n'$.

Now you can substitute for the values of G_n and G_n' and see that we got the formula in the theorem!

The formula we just derived gives new kind of information about the Fibonacci numbers. The first base in the exponential expression is $q_1 = (1+\sqrt{5})/2 \approx 1.618034 > 1$, while the second base q_2 is between -1 and 0. Hence if n increases, then G_n will become very large, while $|G_n'| < \frac{1}{2}$ once $n \geq 2$, and in fact, G_n' becomes very small. This means that
$$F_n \approx G_n = \frac{1}{\sqrt{5}}\left(\frac{1+\sqrt{5}}{2}\right)^n,$$

---

[1] This disappearance of c and n from the equation could be expected. The reason behind it is that if we find a sequence that satisfies Fibonacci's recurrence, then we can multiply its elements by any other real number and get another sequence that satisfies the recurrence. This means that we should not get any condition on c. Further, if we have a sequence that satisfies Fibonacci's recurrence, then starting the sequence anywhere later, it will also satisfy the recurrence. This suggests that we should not get any condition on n.

where the term we ignore is less than $\frac{1}{2}$ if $n \geq 2$ (and tends to 0 if n tends to infinity); this implies that F_n is the integer nearest to G_n.

The base $\tau = (1+\sqrt{5})/2$ is a famous number: It is called the *golden ratio*, and it comes up all over mathematics; for example, it is the ratio between the diagonal and side of a regular pentagon. Another way to characterize it is the following: If $b/a = \tau$, then $(a + b)/b = \tau$. So if the ratio between the longer and shorter sides of a rectangle is τ, then cutting off a square, we are left with a rectangle that is similar to the original.

4.3.2 Define a sequence of integers L_n by $L_1 = 1$, $L_2 = 3$, and $L_{n+1} = L_n + L_{n-1}$. (These numbers are called *Lucas numbers*.) Show that L_n can be expressed in the form $a \cdot q_1^n + b \cdot q_2^n$ (where q_1 and q_2 are the same numbers as in the proof above), and find the values of a and b.

4.3.3 Define a sequence of integers I_n by $I_0 = 0$, $I_1 = 1$, and $I_{n+1} = 4I_n + I_{n-1}$. (a) Find a combinatorial counting problem to which the answer is I_n. (b) Find a formula for I_n.

4.3.4 Alice claims that she knows another formula for the Fibonacci numbers: $F_n = \lceil e^{n/2-1} \rceil$ for $n = 1, 2, \ldots$ (where $e = 2.718281828 \ldots$ is, naturally, the base of the natural logarithm). Is she right?

Review Exercises

4.3.5 In how many ways can you cover a $2 \times n$ chessboard by dominoes?

4.3.6 How many subsets does the set $\{1, 2, \ldots, n\}$ have that contain no two consecutive integers if 0 and n also count as consecutive?

4.3.7 How many subsets does the set $\{1, 2, \ldots, n\}$ have that contain no three consecutive integers? Find a recurrence.

4.3.8 Which number is larger, 2^{100} or F_{100}?

4.3.9 Prove the following identities:

(a) $F_2 + F_4 + F_6 + \cdots + F_{2n} = F_{2n+1} - 1$;

(b) $F_{n+1}^2 - F_n^2 = F_{n-1}F_{n+2}$;

(c) $\binom{n}{0}F_0 + \binom{n}{1}F_1 + \binom{n}{2}F_2 + \cdots + \binom{n}{n}F_n = F_{2n}$;

(d) $\binom{n}{0}F_1 + \binom{n}{1}F_2 + \binom{n}{2}F_3 + \cdots + \binom{n}{n}F_{n+1} = F_{2n+1}$.

4.3.10 Prove (4.4).

4.3.11 Is it true that if F_n is a prime, then n is a prime?

4.3.12 Consider a sequence of numbers $b_0, b_1, b_2, \ldots$ such that $b_0 = 0$, $b_1 = 1$, and $b_2, b_3, \ldots$ are defined by the recurrence

$$b_{k+1} = 3b_k - 2b_{k-1}.$$

Find the value of b_k.

4.3.13 Assume that the sequence $(a_0, a_1, a_2, \ldots)$ satisfies the recurrence

$$a_{n+1} = a_n + 2a_{n-1}.$$

We know that $a_0 = 4$ and $a_2 = 13$. What is a_5?

4.3.14 Recalling the Lucas numbers L_n introduced in Exercise 4.3.2, prove the following identities:

(a) $F_{2n} = F_n L_n$;

(b) $2F_{k+n} = F_k L_n + F_n L_k$;

(c) $2L_{k+n} = 5F_k F_n + L_k L_n$;

(d) $L_{4k} = L_{2k}^2 - 2$;

(e) $L_{4k+2} = L_{2k+1}^2 + 2$.

4.3.15 Prove that if n is a multiple of 4, then F_n is a multiple of 3.

4.3.16   (a) Prove that every positive integer can be written as the sum of different Fibonacci numbers.

(b) Prove even more: every positive integer can be written as the sum of different Fibonacci numbers, so that no two consecutive Fibonacci numbers are used.

(c) Show by an example that the representation in (a) is not unique, but also prove that the more restrictive representation in (b) is.

5
Combinatorial Probability

5.1 Events and Probabilities

Probability theory is one of the most important areas of mathematics from the point of view of applications. In this book we do not attempt to introduce even the most basic notions of probability theory; our only goal is to illustrate the importance of combinatorial results about Pascal's Triangle by explaining a key result in probability theory, the Law of Large Numbers. To do so, we have to talk a little about what probability is.

If we make an observation about our world, or carry out an experiment, the outcome will always depend on chance (to a varying degree). Think of the weather, the stock market, or a medical experiment. Probability theory is a way of modeling this dependence on chance.

We start with making a mental list of all possible outcomes of the experiment (or observation, which we don't need to distinguish). These possible outcomes form a set S. Perhaps the simplest experiment is tossing a coin. This has two outcomes: H (heads) and T (tails). So in this case $S = \{H, T\}$. As another example, the outcomes of throwing a die form the set $S = \{1, 2, 3, 4, 5, 6\}$. In this book we assume that the set $S = \{s_1, s_2, \ldots, s_k\}$ of possible outcomes of our experiment is finite. The set S is often called a *sample space*.

Every subset of S is called an *event* (the event that the observed outcome falls in this subset). So if we are throwing a die, the subset $E = \{2, 4, 6\} \subseteq S$ can be thought of as the event that we throw an even number. Similarly, the subset $L = \{4, 5, 6\} \subseteq S$ corresponds to the event that we throw a number larger than 3.

The intersection of two subsets corresponds to the event that both events occur; for example, the subset $L \cap E = \{4, 6\}$ corresponds to the event that we throw a better-than-average number that is also even. Two events A and B (i.e., two subsets of S) are called *exclusive* if they never occur at the same time, i.e., $A \cap B = \emptyset$. For example, the event $O = \{1, 3, 5\}$ that the outcome of tossing a die is odd and the event E that it is even are exclusive, since $E \cap O = \emptyset$.

5.1.1 What event does the union of two subsets corresponds to?

So let $S = \{s_1, s_2, \dots, s_k\}$ be the set of possible outcomes of an experiment. To get a probability space we assume that each outcome $s_i \in S$ has a "probability" $\mathsf{P}(s_i)$ such that

(a) $\mathsf{P}(s_i) \geq 0$ for all $s_i \in S$,

and

(b) $\mathsf{P}(s_1) + \mathsf{P}(s_2) + \cdots + \mathsf{P}(s_k) = 1$.

Then we call S, together with these probabilities, a *probability space*. For example, if we toss a "fair" coin, then $\mathsf{P}(H) = \mathsf{P}(T) = \frac{1}{2}$. If the dice in our example is of good quality, then we will have $\mathsf{P}(i) = \frac{1}{6}$ for every outcome i.

A probability space in which every outcome has the same probability is called a *uniform probability space*. We shall only discuss uniform spaces here, since they are the easiest to imagine and they are the best for the illustration of combinatorial methods. But you should be warned that in more complicated modeling, nonuniform probability spaces are very often needed. For example, if we are observing whether a day is rainy or not, we will have a 2-element sample space $S = \{\text{RAINY}, \text{NONRAINY}\}$, but these two will typically *not* have the same probability.

The probability of an event $A \subseteq S$ is defined as the sum of probabilities of outcomes in A, and is denoted by $\mathsf{P}(A)$. If the probability space is uniform, then the probability of A is

$$\mathsf{P}(A) = \frac{|A|}{|S|} = \frac{|A|}{k}.$$

5.1.2 Prove that the probability of any event is at most 1.

5.1.3 What is the probability of the event E that we throw an even number with the die? What is the probability of the event $T = \{3, 6\}$ that we toss a number that is divisible by 3?

5.1.4 Prove that if A and B are exclusive, then $\mathsf{P}(A) + \mathsf{P}(B) = \mathsf{P}(A \cup B)$.

5.1.5 Prove that for any two events A and B,

$$\mathsf{P}(A \cap B) + \mathsf{P}(A \cup B) = \mathsf{P}(A) + \mathsf{P}(B).$$

5.2   Independent Repetition of an Experiment

Let us repeat our experiment n times. We can consider this as a single big experiment, and a possible outcome of this repeated experiment is a sequence of length n, consisting of elements of S. Thus the sample space corresponding to this repeated experiment is the set S^n of such sequences. Consequently, the number of outcomes of this "big" experiment is k^n. We consider every sequence equally likely, which means that we consider it a uniform probability space. Thus if $(a_1, a_2, \ldots, a_n)$ is an outcome of the "big" experiment, then we have

$$\mathsf{P}(a_1, a_2, \ldots, a_n) = \frac{1}{k^n}.$$

As an example, consider the experiment of tossing a coin twice. Then $S = \{H, T\}$ (heads, tails) for a single coin toss, and so the sample space for the two coin tosses is $\{HH, HT, TH, TT\}$. The probability of each of these outcomes is $\frac{1}{4}$.

This definition intends to model the situation where the outcome of each repeated experiment is independent of the previous outcomes, in the everyday sense that "there cannot possibly be any measurable influence of one experiment on the other." We cannot go here into the philosophical questions that this notion raises; all we can do is to give a mathematical definition that we can check, using examples, that it correctly expresses the informal notion above.

A key notion in probability is *independence* of events. Informally, this means that information about one event (whether or not it occurred) does not influence the probability of the other. Formally, two events A and B are *independent* if $\mathsf{P}(A \cap B) = \mathsf{P}(A)\mathsf{P}(B)$.

Consider again the experiment of tossing a coin twice. Let A be the event that the first toss is heads; let B be the event that the second toss is heads. Then we have $\mathsf{P}(A) = \mathsf{P}(HH) + \mathsf{P}(HT) = \frac{1}{4} + \frac{1}{4} = \frac{1}{2}$, similarly $\mathsf{P}(B) = \frac{1}{2}$, and $\mathsf{P}(A \cap B) = \mathsf{P}(HH) = \frac{1}{4} = \frac{1}{2} \cdot \frac{1}{2}$. Thus A and B are independent events (as they should be).

As another example, suppose that we toss a coin and simultaneously throw a die. The event H that we toss heads has probability $\frac{1}{2}$. The event K that we see 5 or 6 on the die has probability $\frac{1}{3}$. The event $H \cap K$ that we see heads on the coin and 5 or 6 on the die has probability $\frac{1}{6}$, since out of the 12 possible outcomes (H1, H2, H3, H4, H5, H6, T1, T2, T3, T4, T5, T6) two will have this property. So

$$\mathsf{P}(H \cap K) = \frac{1}{6} = \frac{1}{2} \cdot \frac{1}{3} = \mathsf{P}(H) \cdot \mathsf{P}(E),$$

and thus the events H and K are independent.

Independence of events is a mathematical notion and it does not necessarily mean that they have physically nothing to do with each other. If

$E = \{2, 4, 6\}$ is the event that the result of throwing a dice is even, and $T = \{3, 6\}$ is the event that it is a multiple of 3, then the event E and the event T are independent: we have $E \cap T = \{6\}$ (the only possibility to throw a number that is even and divisible by 3 is to throw 6), and hence

$$P(E \cap T) = \frac{1}{6} = \frac{1}{2} \cdot \frac{1}{3} = P(E)P(T).$$

5.2.1 Which pairs of the events E, O, T, L are independent? Which pairs are exclusive?

5.2.2 Show that $\emptyset$ is independent of every event. Is there any other event with this property?

5.2.3 Consider an experiment with sample space S repeated n times ($n \geq 2$). Let $s \in S$. Let A be the event that the first outcome is s, and let B be the event that the last outcome is s. Prove that A and B are independent.

5.2.4 How many people do you think there are in the world who have the same birthday as their mother? How many people have the same birthday as their mother, father, and spouse?

5.3   The Law of Large Numbers

In this section we study an experiment that consists of n independent coin tosses. For simplicity, assume that n is even, so that $n = 2m$ for some integer m. Every outcome is a sequence of length n, in which each element is either H or T. A typical outcome would look like this:

$$HHTTTHTHTTHTHHHHTHTT$$

(for $n = 20$).

The *Law of Large Numbers* says that if we toss a coin many times, the number of "heads" will be about the same as the number of "tails". How can we make this statement precise? Certainly, this will not *always* be true; one can be extremely lucky or unlucky, and have a winning or loosing streak of arbitrary length. Also, we can't claim that the number of heads is equal to the number of tails; only that they are very likely to be close:

> *Flipping a coin n times, the probability that the percentage of heads is between 49% and 51% tends to 1 as n tends to ∞.*

The statement remains true if we replace 49% by 49.9% and 51% by 50.1%, or indeed by any two numbers strictly less 50% and larger than 50%, respectively. We can state this as a theorem, which is the simplest form of the Law of Large Numbers:

Theorem 5.3.1 *Fix an arbitrarily small positive number ϵ. If we flip a coin n times, the probability that the fraction of heads is between $0.5 - \epsilon$ and $0.5 + \epsilon$ tends to 1 as n tends to ∞.*

This theorem says, for example, that flipping a coin n times, the probability that the number of heads is between 49% and 51% is at least 0.99, if n is large enough. But how large must n be for this to hold? If $n = 49$ (which may sound pretty large) the number of heads can *never* be in this range; there are simply no integers between 49% of 49 (24.01) and 51% of 49 (24.99). How much larger does n have to be to assure that the number of heads is in this range for the majority of outcomes? This is an extremely important question in the statistical analysis of data: we want to know whether a deviation from the expected value is statistically significant.

Fortunately, much more precise formulations of the Law of Large Numbers can be made; one of these we can prove relatively easily, based on what we already know about Pascal's triangle. This proof will show that the Law of Large Numbers is not a mysterious force, but a simple consequence of the properties of binomial coefficients.

Theorem 5.3.2 *Let $0 \le t \le m$. Then the probability that out of $2m$ coin tosses, the number of heads is less than $m - t$ or larger than $m + t$, is at most $e^{-t^2/(m+t)}$.*

To illustrate the power of this theorem, let's go back to our earlier question: *How large should n be in order that the probability that the number of heads is between 49% and 51% is at least 0.99?* We want $m - t$ to be 49% of $n = 2m$, which means that $t = m/50$. The theorem says that the probability that the number of heads is not in this interval is at most $e^{-t^2/(m+t)}$. The exponent here is

$$-\frac{t^2}{m+t} = -\frac{(\frac{m}{50})^2}{m + \frac{m}{50}} = -\frac{m}{2550}.$$

We want $e^{-m/2550} < 0.01$; taking the logarithm and solving for m, we get $m \ge 11744$ suffices. (This is pretty large, but, after all, we are talking about the "Law of Large Numbers.")

Observe that m is in the exponent, so that if m increases, the probability that the number of heads is outside the given interval drops very fast. For example, if $m = 1{,}000{,}000$, then this probability is less than 10^{-170}. Most likely, over the lifetime of the universe it never happens that out of a million coin tosses less than 49% or more than 51% are heads.

Normally, we don't need such a degree of certainty. Suppose that we want to make a claim about the number of heads with 95% certainty, but we would like to narrow the interval into which it falls as much as possible. In other words, we want to choose the smallest possible t so that the probability that the number of heads is less than $m - t$ or larger than $m + t$ less than 0.05. By Theorem 5.3.2, this will be the case if

$$e^{-t^2/(m+t)} < 0.05.$$

(This is only a sufficient condition; if this holds, then the number of heads will be between $m - t$ and $m + t$ with probability at least 0.95. Using more refined formulas, we would find a slightly smaller t that works.) Taking the logarithm, we get

$$-\frac{t^2}{m + t} < -2.996.$$

This leads to a quadratic inequality, which we could solve for t; but it should suffice for this discussion that $t = 2\sqrt{m} + 2$ satisfies it (which is easy to check). So we get an interesting special case:

With probability at least 0.95, the number of heads among $2m$ coin tosses is between $m - 2\sqrt{m} - 2$ and $m + 2\sqrt{m} + 2$.

If m is very large, then $2\sqrt{m} + 2$ is much smaller than m, so we get that the number of heads is very close to m. For example, if $m = 1,000,000$ then $2\sqrt{m} = 2,002 \approx 0.002m$, and so it follows that with probability at least 0.95, the number of heads is within $\frac{1}{5}$ of a percent of $m = n/2$.

It is time now to turn to the proof of Theorem 5.3.2.

Proof. Let A_k denote the event that we toss exactly k heads. It is clear that the events A_k are mutually exclusive. It is also clear that for every outcome of the experiment, exactly one of the A_k occurs.

The number of outcomes for which A_k occurs is the number of sequences of length n consisting of k heads and $n - k$ tails. If we specify which of the n positions are heads, we are done. This can be done in $\binom{n}{k}$ ways, so the set A_k has $\binom{n}{k}$ elements. Since the total number of outcomes is 2^n, we get the following:

$$P(A_k) = \frac{\binom{n}{k}}{2^n}.$$

What is the probability that the number of heads is far from the expected, which is $m = n/2$; say, it is less than $m - t$ or larger than $m + t$, where t is any positive integer not larger than m? Using Exercise 5.1.4, we see that the probability that this happens is

$$\frac{1}{2^{2m}} \left(\binom{2m}{0} + \binom{2m}{1} + \cdots + \binom{2m}{m - t - 1} + \binom{2m}{m + t + 1} + \cdots \right.$$
$$\left. + \binom{2m}{2m - 1} + \binom{2m}{2m} \right).$$

Now we can use Lemma 3.8.2, with $k = m - t$, and get that

$$\binom{2m}{0} + \binom{2m}{1} + \cdots + \binom{2m}{m - t - 1} < 2^{2m-1} \binom{2m}{m - t} \bigg/ \binom{2m}{m}.$$

By (3.9), this can be bounded from above by

$$2^{2m-1}e^{-t^2/(m+t)}.$$

By the symmetry of Pascal's triangle, we also have

$$\binom{2m}{m+t+1} + \cdots + \binom{2m}{2m-1} + \binom{2m}{2m} < 2^{2m}e^{-t^2/(m+t)}.$$

Hence we get that the probability that we toss either fewer than $m - t$ or more than $m + t$ heads is less than $e^{-t^2/(m+t)}$. This proves the theorem. $\square$

5.4 The Law of Small Numbers and the Law of Very Large Numbers

There are two further statistical "laws" (half serious): the *Law of Small Numbers* and the *Law of Very Large Numbers*.

The first one says that if you look at small examples, you can find many strange or interesting patterns that do not generalize to larger numbers. Small numbers exhibit only a small number of patterns, and looking at various properties of small numbers, we are bound to see coincidences. For example, "every odd number is a prime" is true for 3, 5 and 7 (and one may be tempted to say that it is also true for 1, which is even "simpler" than primes: instead of two divisors, it has only one). Of course, this fails for 9.

Primes are strange (as we'll see) and in their irregular sequence, many strange patterns can be observed, which than fail if we move on to larger numbers. A dramatic example is the formula $n^2 - n + 41$. This gives a prime for $n = 0, 1, \ldots, 40$, but for $n = 41$ we get $41^2 - 41 + 41 = 41^2$, which is not a prime.

Fibonacci numbers are not as strange as primes: We have seen many interesting properties of them, and derived an explicit formula in Chapter 4. Still, one can make observations for the beginning of the sequence that do not remain valid if we check them far enough. For example, Exercise 4.3.4 gave a (false) formula for the Fibonacci numbers, namely $\lceil e^{n/2-1} \rceil$, which was correct for the first 10 positive integers n. There are many formulas that give integer sequences, but these sequences can start only so many ways: we are bound to find different sequences that start out the same way.

So the moral of the "Law of Small Numbers" is that to make a mathematical statement, or even to set up a mathematical conjecture, it is not enough to observe some pattern or rule, because you can only observe small instances and there are many coincidences for these. There is nothing wrong with making conjectures in mathematics, generalizing facts observed in special cases, but even a conjecture needs some other justification (an imprecise argument, or a provable special case). A theorem, of course, needs much more: an exact proof.

The Law of Very Large Numbers says that strange coincidences can also be observed if we look at large sets of data. A friend of ours says, "I know two people who were both born on Christmas day. They complain that they get only one set of presents. . . . That's really strange. Are there many more people born on Christmas day than on other days?" No, this is not the explanation. The probability that a person is born on Christmas day is 1/365 (let's ignore leap years), so if you know, say, 400 people, then you can expect 1 or 2 of them to have a birthday on Christmas. Of course, you probably don't remember the birthdays of most people you know; but you are likely to remember those who complain about not getting enough presents!

Would you find it strange, even spooky, if somebody had the same birthday as his/her mother, father, and spouse? But if you have solved Exercise 5.2.4, you know that we have probably about 40 or so such people in the world, and probably a couple of them in the United States.

This is a fertile area for the tabloids and also for believers in the paranormal. We had better leave it at that.

Review Exercises

5.4.1 We throw a die twice. What is the probability that the sum of the points is 8?

5.4.2 Choose an integer uniformly from the set $\{1, 2, 3, \ldots, 30\}$. Let A be the event that it is divisible by 2; let B be the event that it is divisible by 3; let C be the event that it is divisible by 7.

(a) Determine the probabilities of A, B, and C.

(b) Which of the pairs (A, B), (B, C), and (A, C) are independent?

5.4.3 Let A and B be independent events. Express the probability $\mathsf{P}(A \cup B)$ in terms of the probabilities of A and B.

5.4.4 We select a subset X of the set $S = \{1, 2, \ldots, 100\}$ randomly and uniformly (so that every subset has the same probability of being selected). What is the probability that

(a) X has an even number of elements;

(b) both 1 and 100 belong to X;

(c) the largest element of S is 50;

(d) S has at most 2 elements.

5.4.5 We flip a coin n times $(n \geq 1)$. For which values of n are the following pairs of events independent?

(a) The first coin flip was heads; the number of all heads was even.

(b) The first coin flip was head; the number of all heads was more than the number of tails.

(c) The number of heads was even; the number of heads was more than the number of tails.

6
Integers, Divisors, and Primes

In this chapter we discuss properties of integers. This area of mathematics is called *number theory*, and it is a truly venerable field: Its roots go back about 2500 years, to the very beginning of Greek mathematics. One might think that after 2500 years of research, one would know essentially everything about the subject. But we shall see that this is not the case: There are very simple, natural questions that we cannot answer; and there are other simple, natural questions to which an answer has been found only in the last few years!

6.1 Divisibility of Integers

We start with some very basic notions concerning integers. Let a and b be two integers. We say that a *divides* b, or a *is a divisor of* b, or b *is a multiple of* a (these phrases mean the same thing), if there exists an integer m such that $b = am$. In notation: $a \mid b$. If a is not a divisor of b, then we write $a \nmid b$. If $a \neq 0$, then $a \mid b$ means that the ratio b/a is an integer.

If $a \nmid b$ and $a > 0$, then we can still divide b by a with remainder. The remainder r of the division $b \div a$ is an integer that satisfies $0 \leq r < a$. If the quotient of the division with remainder is q, then we have

$$b = aq + r.$$

This is a very useful way of thinking about a division with remainder.

You have probably seen these notions before; the following exercises should help you check whether you remember enough.

6.1.1 Check (using the definition) that $1 \mid a$, $-1 \mid a$, $a \mid a$ and $-a \mid a$ for every integer a.

6.1.2 What does it mean for a, in more everyday terms, if (a) $2 \mid a$; (b) $2 \nmid a$; (c) $0 \mid a$.

6.1.3 Prove that

(a) if $a \mid b$ and $b \mid c$ then $a \mid c$;

(b) if $a \mid b$ and $a \mid c$ then $a \mid b + c$ and $a \mid b - c$;

(c) if $a, b > 0$ and $a \mid b$ then $a \le b$;

(d) if $a \mid b$ and $b \mid a$ then either $a = b$ or $a = -b$.

6.1.4 Let r be the remainder of the division $b \div a$. Assume that $c \mid a$ and $c \mid b$. Prove that $c \mid r$.

6.1.5 Assume that $a \mid b$, and $a, b > 0$. Let r be the remainder of the division $c \div a$, and let s be the remainder of the division $c \div b$. What is the remainder of the division $s \div a$?

6.1.6   (a) Prove that for every integer a, $a - 1 \mid a^2 - 1$.

(b) More generally, for every integer a and positive integer n,

$$a - 1 \mid a^n - 1.$$

6.2   Primes and Their History

An integer $p > 1$ is called a *prime* if it is not divisible by any integer other than $1, -1, p$, and $-p$. Another way of saying this is that an integer $p > 1$ is a prime if it cannot be written as the product of two smaller positive integers. An integer $n > 1$ that is not a prime is called *composite* (the number 1 is considered neither prime nor composite). Thus $2, 3, 5, 7, 11$ are primes, but $4 = 2 \cdot 2$, $6 = 2 \cdot 3$, $8 = 2 \cdot 4$, $9 = 3 \cdot 3$, $10 = 2 \cdot 5$ are not primes. Table 6.1 shows the primes up to 500.

Primes have fascinated people ever since ancient times. Their sequence seems very irregular, yet on closer inspection it seems to carry a lot of hidden structure. The ancient Greeks already knew that there are infinitely many such numbers. (Not only did they know this; they proved it!)

It was not easy to prove any further facts about primes. Their sequence is reasonably smooth, but it has holes and dense spots (see Figure 6.1). How large are these holes? For example, is there a prime number with any given number of digits? The answer to this question will be important for us when we discuss cryptography. The answer is in the affirmative, but this fact was not proved until the mid-nineteenth century, and many similar questions are open even today.

1, *2*, *3*, 4, *5*, 6, *7*, 8, 9, 10, *11*, 12, *13*, 14, 15, 16, *17*, 18, *19*, 20, 21, 22, *23*, 24, 25, 26, 27, 28, *29*, 30, *31*, 32, 33, 34, 35, 36, *37*, 38, 39, 40, *41*, 42, *43*, 44, 45, 46, *47*, 48, 49, 50, 51, 52, *53*, 54, 55, 56, 57, 58, *59*, 60, *61*, 62, 63, 64, 65, 66, *67*, 68, 69, 70, *71*, 72, *73*, 74, 75, 76, 77, 78, *79*, 80, 81, 82, *83*, 84, 85, 86, 87, 88, *89*, 90, 91, 92, 93, 94, 95, 96, *97*, 98, 99, 100, *101*, 102, *103*, 104, 105, 106, *107*, 108, *109*, 110, 111, 112, *113*, 114, 115, 116, 117, 118, 119, 120, 121, 122, 123, 124, 125, 126, *127*, 128, 129, 130, *131*, 132, 133, 134, 135, 136, *137*, 138, *139*, 140, 141, 142, 143, 144, 145, 146, 147, 148, *149*, 150, *151*, 152, 153, 154, 155, 156, *157*, 158, 159, 160, 161, 162, *163*, 164, 165, 166, 167, 168, 169, 170, 171, 172, *173*, 174, 175, 176, 177, 178, *179*, 180, *181*, 182, 183, 184, 185, 186, 187, 188, 189, 190, *191*, 192, *193*, 194, 195, 196, *197*, 198, *199*, 200, 201, 202, 203, 204, 205, 206, 207, 208, 209, 210, *211*, 212, 213, 214, 215, 216, 217, 218, 219, 220, 221, 222, *223*, 224, 225, 226, *227*, 228, *229*, 230, 231, 232, 233, 234, 235, 236, 237, 238, *239*, 240, *241*, 242, 243, 244, 245, 246, 247, 248, 249, 250, *251*, 252, 253, 254, 255, 256, *257*, 258, 259, 260, 261, 262, *263*, 264, 265, 266, 267, 268, *269*, 270, *271*, 272, 273, 274, 275, 276, *277*, 278, 279, 280, *281*, 282, *283*, 284, 285, 286, 287, 288, 289, 290, 291, 292, *293*, 294, 295, 296, 297, 298, 299, 300, 301, 302, 303, 304, 305, 306, *307*, 308, 309, 310, *311*, 312, *313*, 314, 315, 316, *317*, 318, 319, 320, 321, 322, 323, 324, 325, 326, 327, 328, 329, 330, *331*, 332, 333, 334, 335, 336, *337*, 338, 339, 340, 341, 342, 343, 344, 345, 346, *347*, 348, *349*, 350, 351, 352, *353*, 354, 355, 356, 357, 358, *359*, 360, 361, 362, 363, 364, 365, 366, *367*, 368, 369, 370, 371, 372, *373*, 374, 375, 376, 377, 378, *379*, 380, 381, 382, *383*, 384, 385, 386, 387, 388, *389*, 390, 391, 392, 393, 394, 395, 396, *397*, 398, 399, 400, *401*, 402, 403, 404, 405, 406, 407, 408, *409*, 410, 411, 412, 413, 414, 415, 416, 417, 418, *419*, 420, *421*, 422, 423, 424, 425, 426, 427, 428, 429, 430, *431*, 432, *433*, 434, 435, 436, 437, 438, *439*, 440, 441, 442, *443*, 444, 445, 446, 447, 448, *449*, 450, 451, 452, 453, 454, 455, 456, *457*, 458, 459, 460, *461*, 462, *463*, 464, 465, 466, *467*, 468, 469, 470, 471, 472, 473, 474, 475, 476, 477, 478, *479*, 480, 481, 482, 483, 484, 485, 486, *487*, 488, 489, 490, *491*, 492, 493, 494, 495, 496, 497, 498, *499*, 500

TABLE 6.1. The primes up to 500.

A new wave of developments in the theory of prime numbers came with the spread of computers. How do you decide about a positive integer n whether it is a prime? Surely, this is a finite problem (you can try out all smaller positive integers to see whether any of them is a proper divisor), but such simple methods become impractical as soon as the number of digits is more than 20 or so.

It is only 25 years since much more efficient algorithms (computer programs) have existed to test whether a given integer is a prime. We will get a glimpse of these methods later. Using these methods, one can now rather easily determine about a number with 1000 digits whether it is a prime or not.

If an integer larger than 1 is not itself a prime, then it can be written as a product of primes: We can write it as a product of two smaller positive integers; if one of these is not a prime, we write it as the product of two smaller integers, etc; sooner or later we must end up with only primes. The ancient Greeks also knew (and proved!) a subtler fact about this represen-

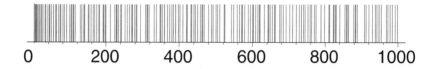

FIGURE 6.1. A bar chart of primes up to 1000.

tation, that *it is unique*. What this means is that there is no other way of writing n as a product of primes (except, of course, we can multiply the same primes in a different order). To prove this takes some sophistication (as we'll see in the next section), and to recognize the necessity of such a result was quite an accomplishment; but this is all more than 2000 years old!

It is really surprising that, even today, no efficient way is known to *find* such a decomposition. Of course, powerful supercomputers and massively parallel systems can be used to find decompositions by brute force for fairly large numbers; the current record is around 140 digits, and the difficulty grows very fast (exponentially) with the number of digits. To find the prime decomposition of a given number with 400 digits, by any of the known methods, is way beyond the possibilities of computers in the foreseeable future.

6.3   Factorization into Primes

We have seen that every integer larger than 1 that is not a prime itself can be written as a product of primes. We can even say that *every* positive integer can be written as a product of primes: Primes can be considered as "products with one factor," and if you wish, the integer 1 can be thought of as the "empty product." With this in mind, we can state and prove the following theorem, announced above, sometimes called the "Fundamental Theorem of Arithmetic".

Theorem 6.3.1 *Every positive integer can be written as the product of primes, and this factorization is unique up to the order of the prime factors.*

Proof. We prove this theorem by a version of induction, which is sometimes called the "minimal criminal" argument. The proof is indirect: we suppose that the assertion is false, and using this assumption, we derive a logical contradiction.

So assume that there exists an integer with two different factorizations; call such an integer a "criminal." There may be many criminals, but we consider the *smallest* one. Being a criminal, this has at least two different factorizations:

$$n = p_1 \cdot p_2 \cdots p_m = q_1 \cdot q_2 \cdots q_k .$$

We may assume that p_1 is the smallest prime occurring in these factorizations. (Indeed, if necessary, we can interchange the left-hand side and the right-hand side so that the smallest prime in any of the two factorizations occurs on the left-hand side; and then change the order of the factors on the left-hand side so that the smallest factor comes first. In the usual slang of mathematics, we say that we may assume that p_1 is the smallest prime *without loss of generality*.) We are going to produce a smaller criminal; this will be a contradiction, since we assumed that n was the smallest one.

The number p_1 cannot occur among the factors q_i; otherwise, we can divide both sides by p_1 and get a smaller criminal.

Divide each q_i by p_1 with residue: $q_i = p_1 a_i + r_i$, where $0 \le r_i < p_1$. We know that $r_i \ne 0$, since a prime cannot be a divisor of another prime.

Let $n' = r_1 r_2 \cdots r_k$. We show that n' is a smaller criminal. Trivially $r_i < p_1 < q_i$, and so $n' = r_1 r_2 \cdots r_k < q_1 q_2 \cdots q_k = n$. We show that n', too, has two different factorizations into primes. One of these can be obtained from the definition $n' = r_1 r_2 \cdots r_k$. Here the factors may not be primes, but we can break them down into products of primes, so that we end up with a decomposition of n'.

To get another decomposition, we observe that $p_1 \mid n'$. Indeed, we can write the definition of n' in the form

$$n' = (q_1 - a_1 p_1)(q_2 - a_2 p_1) \cdots (q_k - a_k p_1),$$

and if we expand, then every term will be divisible by p_1. (One of the terms is $q_1 q_2 \cdots q_k$, which is equal to n and so divisible by p_1. All the other terms contain p_1 as a factor.) Now we divide n' by p_1 and then continue to factor n'/p_1, to get a factorization of n'.

But are these two factorizations different? Yes! The prime p_1 occurs in the second, but it cannot occur in the first, where every prime factor is smaller than p_1.

Thus we have found a smaller criminal. Since n was supposed to be the smallest among all criminals, this is a contradiction. The only way to resolve this contradiction is to conclude that there are no criminals; our "indirect assumption" was false, and no integer can have two different prime factorizations. □

6.3.1 Read carefully the following "minimal criminal" argument:

ASSERTION. *Every negative integer is odd.*

PROOF. Suppose, by way of contradiction, that there are negative integers that are even. Call these integers criminals, and let n be a minimal criminal. Consider the number $2n$. This is smaller than n (recall that n is negative!), so it is a smaller criminal. But we assumed that n was the smallest criminal, so this is a contradiction.

This assertion is obviously wrong. Where is the error in the proof?

As an application of Theorem 6.3.1, we prove a fact that was known to the Pythagoreans (students of the great Greek mathematician and philosopher Pythagoras) in the sixth century B.C.

Theorem 6.3.2 *The number $\sqrt{2}$ is irrational.*

(A real number is *irrational* if it cannot be written as the ratio of two integers. For the Pythagoreans, the question arose from geometry: They wanted to know whether the diagonal of a square is "commeasurable" with its side, that is, whether there is any segment that is contained in both of them an integer number of times. The above theorem answered this question in the negative, causing a substantial turmoil in their ranks.)

Proof. We give an indirect proof again: We suppose that $\sqrt{2}$ is rational, and derive a contradiction. What the indirect assumption means is that $\sqrt{2}$ can be written as the quotient of two positive integers: $\sqrt{2} = a/b$. Squaring both sides and rearranging, we get $2b^2 = a^2$.

Now consider the prime factorization of both sides, and in particular, the prime number 2 on both sides. Suppose that 2 occurs m times in the prime factorization of a and n times in the prime factorization of b. Then it occurs $2m$ times in the prime factorization of a^2. On the other hand, it occurs $2n$ times in the prime factorization of b^2, and thus it occurs $2n + 1$ times in the prime factorization of $2b^2$. Since $2b^2 = a^2$, and the prime factorization is unique, we must have $2n + 1 = 2m$. But this is impossible, since $2n + 1$ is odd but $2m$ is even. This contradiction proves that $\sqrt{2}$ must be irrational. □

6.3.2 Are there any even primes?

6.3.3   (a) Prove that if p is a prime, a and b are integers, and $p \mid ab$, then either $p \mid a$ or $p \mid b$ (or both).

  (b) Suppose that a and b are integers and $a \mid b$. Also suppose that p is a prime and $p \mid b$ but $p \nmid a$. Prove that p is a divisor of the ratio b/a.

6.3.4 Prove that the prime factorization of a number n contains at most $\log_2 n$ factors.

6.3.5 Let p be a prime and $1 \leq a \leq p - 1$. Consider the numbers $a, 2a, 3a, \ldots, (p - 1)a$. Divide each of them by p, to get residues $r_1, r_2, \ldots, r_{p-1}$. Prove that every integer from 1 to $p-1$ occurs *exactly once* among these residues.
[Hint: First prove that no residue can occur twice.]

6.3.6 Prove that if p is a prime, then $\sqrt{p}$ is irrational. More generally, prove that if n is an integer that is not a square, then $\sqrt{n}$ is irrational.

6.3.7 Try to formulate and prove an even more general theorem about the irrationality of the numbers $\sqrt[k]{n}$.

6.4   On the Set of Primes

The following theorem was known to Euclid in the third century B.C.

Theorem 6.4.1 *There are infinitely many primes.*

Proof. What we need to do is to show that for every positive integer n, there is a prime number larger than n. To this end, consider the number $n! + 1$, and any prime divisor p of it. We show that $p > n$. Again, we use an indirect proof, supposing that $p \leq n$ and deriving a contradiction. If $p \leq n$ then $p \mid n!$, since it is one of the integers whose product is $n!$. We also know that $p \mid n! + 1$, and so p is a divisor of the difference $(n! + 1) - n! = 1$. But this is impossible, and thus p must be larger than n. $\qquad\square$

If we look at various charts or tables of primes, our main impression is that there is a lot of irregularity in them. For example, Figure 6.1 represents each prime up to 1000 by a bar. We see large "gaps", and then we also see primes that are very close. We can prove that these gaps get larger and larger as we consider larger and larger numbers; somewhere out there is a string of 100 consecutive composite numbers; somewhere (much farther away) there is a string of 1000 consecutive composite numbers, etc. To state this in a mathematical form:

Theorem 6.4.2 *For every positive integer k, there exist k consecutive composite integers.*

Proof. We can prove this theorem by an argument quite similar to the proof of Theorem 6.4.1. Let $n = k + 1$ and consider the numbers

$$n! + 2, \ n! + 3, \ \ldots, \ n! + n.$$

Can any of these be a prime? The answer is no: The first number is even, since $n!$ and 2 are both even. The second number is divisible by 3, since $n!$ and 3 are both divisible by 3 (assuming that $n > 2$). In general $n! + i$ is divisible by i, for every $i = 2, 3, \ldots, n$. Hence these numbers cannot be primes, and so we have found $n - 1 = k$ consecutive composite numbers. $\square$

What about the opposite question, finding primes very close to each other? Since all primes except 2 are odd, the difference between two primes must be at least two, except for 2 and 3. Two primes whose difference is 2 are called *twin primes*. Thus $(3,5)$, $(5,7)$, $(11,13)$, $(17,19)$ are twin primes. Looking at the table of the primes up to 500, we find many twin primes; extensive computation shows that there are twin primes with hundreds of digits. However, it is not known whether there are infinitely many twin primes! (Almost certainly there are, but no proof of this fact has been found, in spite of the efforts of many mathematicians for over 2000 years!)

Another way of turning Theorem 6.4.2 around is to ask, how large can these gaps be, relative to where they are on the number line? Could it happen that there is no prime at all with, say, 100 digits? This is again a very difficult question, but here we do know the answer. (No, this does not happen.)

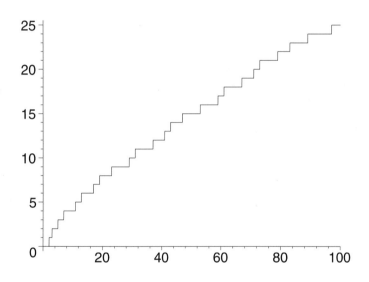

FIGURE 6.2. The graph of $\pi(n)$ from 1 to 100.

One of the most important questions about primes is, how many primes are there up to a given number n? We denote the number of primes up to n by $\pi(n)$. Figure 6.2 illustrates the graph of this function in the range 1 to 100, and Figure 6.3 shows the range 1 to 2000. We can see that the function grows reasonably smoothly, and that its slope decreases slowly. An exact formula for $\pi(n)$ is certainly impossible to obtain. Around 1900, a powerful result called the *Prime Number Theorem* was proved by Hadamard and de la Vallée Poussin.

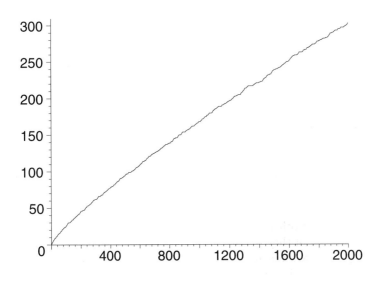

FIGURE 6.3. The graph of $\pi(n)$ from 1 to 2000.

Theorem 6.4.3 (The Prime Number Theorem) *Let $\pi(n)$ denote the number of primes among $1, 2, \ldots, n$. Then*

$$\pi(n) \sim \frac{n}{\ln n}.$$

(Here $\ln n$ means the "natural logarithm," i.e., to logarithm to the base $e = 2.718281\ldots$. Also recall that the notation means that the quotient

$$\pi(n) \Big/ \frac{n}{\ln n}$$

will be arbitrarily close to 1 as n gets large.)

The proof of the Prime Number Theorem is very difficult; the fact that the number of primes up to n is about $n/\ln n$ was observed empirically in the eighteenth century, but it took more than 100 years until Hadamard and de la Vallée Poussin proved it in 1896.

As an illustration of the use of this theorem, let us find the answer to a question that we have posed in the introduction: How many primes with (say) 200 digits are there? We get the answer by subtracting the number of primes up to 10^{199} from the number of primes up to 10^{200}. By the Prime Number Theorem, this number is about

$$\frac{10^{200}}{200 \ln 10} - \frac{10^{199}}{199 \ln 10} \approx 1.95 \cdot 10^{197}.$$

This is a lot of primes! Comparing this with the total number of positive integers with 200 digits, which we know is $10^{200} - 10^{199} = 9 \cdot 10^{199}$, we get

$$\frac{9 \cdot 10^{199}}{1.95 \cdot 10^{197}} \approx 460.$$

Thus among the integers with 200 digits, about one in every 460 is a prime.

(Warning: This argument is not precise; the main source of concern is that in the Prime Number Theorem we stated only that $\pi(n)$ is close to $n/\ln n$ if n is sufficiently large. One can say more about how large n has to be to have, say, an error less than 1 percent, but this leads to even more difficult questions, which are even today not completely resolved.)

There are many other simple observations one can make by looking at tables of primes, but they tend to be very difficult, and most of them are not resolved even today, in some cases after 2,500 years of attempts. We have already mentioned the problem of twin primes. Another famous unsolved problem is *Goldbach's conjecture*. This states that *every even integer larger than 2 can be written as the sum of two primes*. Goldbach also formulated a conjecture about odd numbers: *Every odd integer larger than 5 can be written as the sum of three primes*. This second conjecture was essentially proved, using very deep methods, by Vinogradov in the 1930s. We said "essentially" since the proof works only for numbers that are very large, and the possibility of a finite number of exceptions remains open.

Suppose that we have an integer n and want to know how soon after n we can be sure of finding a prime. For example, how small, or large, is the first prime with at least 100 digits? Our proof of the infinitude of primes gives that for every n there is a prime between n and $n! + 1$. This is a very week statement; it says, for example, that there is a prime between 10 and $10! + 1 = 3,628,801$, while of course the next prime is 11. The Russian mathematician P.L. Chebyshev proved in the mid-nineteenth century that there is always a prime between n and $2n$. It has now been proved that there is always a prime between two consecutive cubes (say, between $10^3 = 1000$ and $11^3 = 1331$). But it is another old, famous, and still unsolved problem whether there is always a prime between two consecutive squares. (Try this out: you'll in fact find many primes. For example, between $100 = 10^2$ and $121 = 11^2$ we find 101, 103, 107, 109, 113. Between $100^2 = 10,000$ and $101^2 = 10,201$ we find 10,007, 10,009, 10,037, 10,039, 10,061, 10,067, 10,069, 10,079, 10,091, 10,093, 10,099, 10,103, 10,111, 10,133, 10,139, 10,141, 10,151, 10,159, 10,163, 10,169, 10,177, 10,181, 10,193.)

6.4.1 Show that among k-digit numbers, one in about every $2.3k$ is a prime.

Pierre de Fermat

6.5 Fermat's "Little" Theorem

Primes are important because we can compose every integer from them; but it turns out that they also have many other, often surprising, properties. One of these was discovered by the French mathematician Pierre de Fermat (1601–1655), now called the "Little" Theorem of Fermat.

Theorem 6.5.1 (Fermat's Theorem) *If p is a prime and a is an integer, then $p \mid a^p - a$.*

Before proving this theorem, we remark that it is often stated in the following form: *If p is a prime and a is an integer not divisible by p, then*

$$p \mid a^{p-1} - 1. \tag{6.1}$$

The fact that these two assertions are equivalent (in the sense that if we know the truth of one, it is easy to prove the other) is left to the reader as Exercise 6.10.20.

To prove Fermat's Theorem, we need a lemma, which states another divisibility property of primes (but is easier to prove):

Lemma 6.5.2 *If p is a prime and $0 < k < p$, then $p \mid \binom{p}{k}$.*

Proof. We know by Theorem 1.8.1 that

$$\binom{p}{k} = \frac{p(p-1)\cdots(p-k+1)}{k(k-1)\cdots 1}.$$

Here p divides the numerator, but not the denominator, since all factors in the denominator are smaller than p, and we know by Exercise 6.3.3(a) that if a prime p does not divide any of these factors, then it does not divide the product. Hence it follows (see Exercise 6.3.3(b)) that p is a divisor of $\binom{p}{k}$. $\square$

Proof [of Theorem 6.5.1]. Now we can prove Fermat's Theorem by induction on a. The assertion is trivially true if $a = 0$. Let $a > 0$, and write $a = b + 1$. Then

$$a^p - a = (b+1)^p - (b+1)$$

$$= b^p + \binom{p}{1}b^{p-1} + \cdots + \binom{p}{p-1}b + 1 - b - 1$$

$$= (b^p - b) + \binom{p}{1}b^{p-1} + \cdots + \binom{p}{p-1}b.$$

Here the expression $b^p - b$ in parenthesis is divisible by p by the induction hypothesis, while the other terms are divisible by p by lemma 6.5.2. It follows that $a^p - a$ is also divisible by p, which completes the induction. $\square$

Let us make a remark about the history of mathematics here. Fermat is most famous for his "last" Theorem, which is the following assertion:

> *If $n > 2$, then the sum of the nth powers of two positive integers is never the nth power of a positive integer.*

(The assumption that $n > 2$ is essential: There are examples of two squares whose sum is a third square. For example, $3^2 + 4^2 = 5^2$, or $5^2 + 12^2 = 13^2$. In fact, there are infinitely many such triples of squares, see Exercise 6.6.7.)

Fermat claimed in a note that he had proved this, but never wrote down the proof. This statement remained perhaps the most famous unsolved problem in mathematics until 1995, when Andrew Wiles (in one part with the help of Robert Taylor) finally proved it.

6.5.1 Show by examples that neither the assertion in lemma 6.5.2 nor Fermat's "Little" Theorem remains valid if we drop the assumption that p is a prime.

6.5.2 Consider a regular p-gon, and for a fixed k $(1 \le k \le p - 1)$, consider all k-subsets of the set of its vertices. Put all these k-subsets into a number of boxes: We put two k-subsets into the same box if they can be rotated into each other. For example, all k-subsets consisting of k consecutive vertices will belong to one and the same box.

(a) Prove that if p is a prime, then each box will contain exactly p of these rotated copies.

(b) Show by an example that (a) does not remain true if we drop the assumption that p is a prime.

(c) Use (a) to give a new proof of Lemma 6.5.2.

6.5.3 Imagine numbers written in base a, with at most p digits. Put two numbers in the same box if they arise by a cyclic shift from each other. How many will be in each class? Give a new proof of Fermat's Theorem this way.

6.5.4 Give a third proof of Fermat's "Little" Theorem based on Exercise 6.3.5. [Hint: Consider the product $a(2a)(3a)\cdots((p-1)a)$.]

6.6   The Euclidean Algorithm

So far, we have discussed several notions and results concerning integers. Now we turn our attention to the question of how to do computations in connection with these results. How do we decide whether or not a given number is a prime? How do we find the prime factorization of a number?

We can do basic arithmetic—addition, subtraction, multiplication, division with remainder—efficiently, and we will not discuss this here.

The key to a more advanced algorithmic number theory is an algorithm that computes the *greatest common divisor* of two positive integers a and b. This is defined as the largest positive integer that is a divisor of both a and b. (Since 1 is always a common divisor, and no common divisor is larger than either integer, this definition makes sense: There is always at least one common divisor, and in the set of common divisors there must be a greatest element.) The greatest common divisor of a and b is denoted by $\gcd(a, b)$. Thus

$$\gcd(1,6) = 1, \qquad \gcd(2,6) = 2, \qquad \gcd(3,6) = 3,$$

$$\gcd(4,6) = 2, \qquad \gcd(5,6) = 1, \qquad \gcd(6,6) = 6.$$

We say that two integers are *relatively prime* if their greatest common divisor is 1. It will be convenient to define $\gcd(a, 0) = a$ for every $a \geq 0$.

A somewhat similar notion is the *least common multiple* of two integers, which is the least positive integer that is a multiple of both integers. It is denoted by $\mathrm{lcm}(a, b)$. For example,

$$\mathrm{lcm}(1,6) = 6, \qquad \mathrm{lcm}(2,6) = 6, \qquad \mathrm{lcm}(3,6) = 6,$$

$$\mathrm{lcm}(4,6) = 12, \qquad \mathrm{lcm}(5,6) = 30, \qquad \mathrm{lcm}(6,6) = 6.$$

The greatest common divisor of two positive integers can be found quite simply by using their prime factorizations: Look at the common prime factors, raise each to the smaller of the two exponents, and take the product of these prime powers. For example, $900 = 2^2 \cdot 3^2 \cdot 5^2$ and $54 = 2 \cdot 3^3$, and hence $\gcd(900, 54) = 2 \cdot 3^2 = 18$.

The trouble with this method is that it is very difficult to find the prime factorization of large integers. The algorithm to be discussed in this section will compute the greatest common divisor of two integers in a much faster way, without finding their prime factorizations. This algorithm is an important ingredient of almost all other algorithms involving computation with integers. (And, as we see it from its name, it goes back to the great Greek mathematician Euclid!)

6.6.1 Show that if a and b are positive integers with $a \mid b$, then $\gcd(a, b) = a$.

6.6.2   (a) Prove that $\gcd(a, b) = \gcd(a, b - a)$.

(b) Let r be the remainder if we divide b by a. Then $\gcd(a, b) = \gcd(a, r)$.

6.6.3   (a) If a is even and b is odd, then $\gcd(a, b) = \gcd(a/2, b)$.

(b) If both a and b are even, then $\gcd(a, b) = 2\gcd(a/2, b/2)$.

6.6.4 How can you express the least common multiple of two integers if you know the prime factorization of each?

6.6.5 Suppose that you are given two integers, and you know the prime factorization of one of them. Describe a way of computing the greatest common divisor of these numbers.

6.6.6 Prove that for any two integers a and b,

$$\gcd(a, b)\operatorname{lcm}(a, b) = ab.$$

6.6.7 Three integers a, b, and c form a *Pythagorean triple* if $a^2 + b^2 = c^2$.

(a) Choose any three integers x, y, and z, and let $a = 2xyz$, $b = (x^2 - y^2)z$, $c = (x^2 + y^2)z$. Check that (a, b, c) is a Pythagorean triple.

(b) Prove that all Pythagorean triples arise this way: If a, b, c are integers such that $a^2 + b^2 = c^2$, then there are other integers x, y, and z such that a, b, and c can be expressed by the formulas above.

[Hint: First, show that the problem can be reduced to the case where $\gcd(a, b, c) = 1$, a is even, and b, c are odd. Second, write $a^2 = (b-c)(b+c)$ and use this to argue that $(b + c)/2$ and $b - c)/2$ are squares.]

Now we turn to the Euclidean Algorithm. It is based on two simple facts, already familiar as Exercises 6.6.1 and 6.6.2.

Suppose that we are given two positive integers a and b, and we want to find their greatest common divisor. Here is what we do:

1. If $a > b$ then we interchange a and b.

2. If $a > 0$, divide b by a, to get a remainder r. Replace b by r and return to 1.

3. Else (if $a = 0$), return b as the g.c.d. and halt.

When you carry out the algorithm, especially by hand, there is no reason to interchange a and b if $a < b$: we can simply divide the larger number by the smaller (with remainder), and replace the larger number by the remainder if the remainder is not 0. Let us do some examples.

$$\gcd(300, 18) = \gcd(12, 18) = \gcd(12, 6) = 6.$$
$$\gcd(101, 100) = \gcd(1, 100) = 1.$$
$$\gcd(89, 55) = \gcd(34, 55) = \gcd(34, 21) = \gcd(13, 21) = \gcd(13, 8)$$
$$= \gcd(5, 8) = \gcd(5, 3) = \gcd(2, 3) = \gcd(2, 1) = 1.$$

You can check in each case (using the prime factorization of the numbers) that the result is indeed the g.c.d.

If we describe an algorithm, the first thing to worry about is whether it terminates at all. So why is the Euclidean Algorithm finite? This is easy: The numbers never increase, one of them decreases whenever step 2 is executed, and the remain nonnegative; so the whole procedure cannot last infinitely long.

Then, of course, we have to make sure that our algorithm yields what we need. This is clear: Step 1 (interchanging the numbers) trivially does not change the greatest common divisor; step 3 (replacing the larger number by the remainder of a division) does not change the greatest common divisor by Exercise 6.6.2(b). And when we halt at step 2, the number returned is indeed the greatest common divisor of the two current numbers by Exercise 6.6.1.

A third, and more subtle, question you should ask when designing an algorithm: How long does it take? How many steps will it make before it terminates? We can get a bound from the argument that proves finite termination: Since one or the other number decreases any time the loop of steps 1 and 2 is executed, it will certainly halt in fewer than $a + b$ iterations. This is really not a great time bound: If we apply the Euclidean Algorithm to two numbers with 100 digits, then this bound of $a + b$ says that it will not last longer than $2 \cdot 10^{100}$ steps, which is an astronomical number, and therefore useless. But luckily this is only an upper bound, and a most pessimistic one at that; the examples we considered seem to show that the algorithm terminates much faster than this.

But the examples also suggest that this question is quite delicate. We see that the Euclidean Algorithm may be quite different in length, depending on the numbers in question. Some of the possible observations made from these examples are contained in the following exercises.

6.6.8 Show that the Euclidean Algorithm can terminate in two steps for arbitrarily large positive integers, even if their g.c.d. is 1.

6.6.9 Describe the Euclidean Algorithm applied to two consecutive Fibonacci numbers. Use your description to show that the Euclidean Algorithm can take arbitrarily many steps.

So what *can* we say about how long the Euclidean Algorithm lasts? The key to the answer is the following lemma:

Lemma 6.6.1 *During the execution of the Euclidean Algorithm, the product of the two current numbers drops by a factor of at least 2 in each iteration.*

Proof. To see that this is so, consider the step where the pair (a, b) $(a < b)$ is replaced by the pair (r, a) (recall that r is the remainder of b when divided by a). Then we have $r < a$ and $a + r \leq b$. Hence $b \geq a + r > 2r$, and so $ar < \frac{1}{2}ab$ as claimed. □

Suppose that we apply the Euclidean Algorithm to two numbers a and b and we make k steps. It follows by Lemma 6.6.1 that after the k steps, the product of the two current numbers will be at most $ab/2^k$. Since this is a positive integer and so at least 1, we get that

$$ab \geq 2^k,$$

and hence

$$k \leq \log_2(ab) = \log_2 a + \log_2 b.$$

Thus we have proved the following.

Theorem 6.6.2 *The number of steps of the Euclidean Algorithm applied to two positive integers a and b is at most $\log_2 a + \log_2 b$.*

We have replaced the sum of the numbers by the sum of the logarithms of the numbers in the bound on the number of steps, which is a really substantial improvement. For example, the number of iterations in computing the g.c.d. of two 300-digit integers is less than $2 \log_2 10^{300} = 600 \log_2 10 < 2000$. Quite a bit less than $2 \cdot 10^{300}$, which was our first naive estimate! Note that $\log_2 a$ is less than the number of bits of a (when written in base 2), so we can say that the Euclidean Algorithm does not take more iterations than the number of bits needed to write down the numbers in base 2.

The theorem above gives only an upper bound on the number of steps the Euclidean Algorithm takes; we can be much luckier. For example, when we apply the Euclidean Algorithm to two consecutive integers, it takes only one step. But sometimes, one cannot do much better. If you did exercise 6.6.9, you saw that when applied to two consecutive Fibonacci numbers F_k and F_{k+1}, the Euclidean Algorithm takes $k - 1$ steps. On the other hand, the lemma above gives the bound

$$\log_2 F_k + \log_2 F_{k+1} \approx \log_2 \left(\frac{1}{\sqrt{5}} \left(\frac{1+\sqrt{5}}{2} \right)^k \right) + \log_2 \left(\frac{1}{\sqrt{5}} \left(\frac{1+\sqrt{5}}{2} \right)^{k+1} \right)$$

$$= -\log_2 5 + (2k+1) \log_2 \left(\frac{1+\sqrt{5}}{2} \right) \approx 1.388k - 1.628,$$

so we have overestimated the number of steps only by a factor of about 1.388, or less than 40%.

Fibonacci numbers are not only good for giving examples of large numbers for which we can see how the Euclidean Algorithm works; they are also useful in obtaining an even better bound on the number of steps. We state the result as an exercise. Its content is that, in a sense, the Euclidean Algorithm is longest on two consecutive Fibonacci numbers.

6.6.10 Suppose that $a < b$ and the Euclidean Algorithm applied to a and b takes k steps. Prove that $a \geq F_k$ and $b \geq F_{k+1}$.

6.6.11 Consider the following version of the Euclidean Algorithm to compute $\gcd(a,b)$: (1) Swap the numbers if necessary to have $a \leq b$; (2) if $a = 0$, then return b; (3) if $a \neq 0$, then replace b by $b - a$ and go to (1).

(a) Carry out this algorithm to compute $\gcd(19, 2)$.

(b) Show that the modified Euclidean Algorithm always terminates with the right answer.

(c) How long does this algorithm take, in the worst case, when applied to two 100-digit integers?

6.6.12 Consider the following version of the Euclidean Algorithm to compute $\gcd(a,b)$. Start with computing the largest power of 2 dividing both a and b. If this is 2^r, then divide a and b by 2^r. After this "preprocessing," do the following:

(1) Swap the numbers if necessary to have $a \leq b$.

(2) If $a \neq 0$, then check the parities of a and b; if a is even, and b is odd, then replace a by $a/2$; if both a and b are odd, then replace b by $b - a$; in each case, go to (1).

(3) if $a = 0$, then return $2^r b$ as the g.c.d.

Now come the exercises:

(a) Carry out this algorithm to compute $\gcd(19, 2)$.

(b) It seems that in step (2), we ignored the case where both a and b are even. Show that this never occurs.

(c) Show that the modified Euclidean Algorithm always terminates with the right answer.

(d) Show that this algorithm, when applied to two 100-digit integers, does not take more than 1500 iterations.

The Euclidean Algorithm gives much more than just the greatest common divisor of the two numbers. The main observation is that if we carry out the Euclidean Algorithm to compute the greatest common divisor of two positive integers a and b, all the numbers we produce during the computation can be written as the sum of an integer multiple of a and an integer multiple of b.

As an example, let's recall the computation of $\gcd(300, 18)$:

$$\gcd(300, 18) = \gcd(12, 18) = \gcd(12, 6) = 6.$$

Here the number 12 was obtained as the remainder of the division $300 \div 18$; this means that it was obtained by subtracting from 300 the highest multiple of 18 that is smaller that 300: $12 = 300 - 16 \cdot 18$. Let's record it in this form:

$$\gcd(300, 18) = \gcd(300 - 16 \cdot 18, 18).$$

Next, we obtained 6 by subtracting 12 from 18, which we can do so that we maintain the form of (multiple of 300)+(multiple of 18):

$$\gcd(300 - 16 \cdot 18, 18) = \gcd(300 - 16 \cdot 18, 17 \cdot 18 - 300).$$

So it follows that the g.c.d. itself, namely 6, is of this form:

$$6 = 17 \cdot 18 - 300.$$

Let us prove formally that all the numbers produced by the Euclidean Algorithm for $\gcd(a, b)$ can be written as the sum of an integer multiple of a and an integer multiple of b. Suppose that this holds for two consecutive numbers we computed, so that one is $a' = am + bn$, and the other is $b' = ak + bl$, where m, n, k, l are integers (not necessarily positive). Then in the next step we compute (say) the remainder of b' modulo a', which is

$$a' - qb' = (am + bn) - q(ak + bl) = a(m - qk) + b(n - ql),$$

which is of the right form again.

In particular, we get the following:

Theorem 6.6.3 *Let $d = \gcd(a, b)$. Then d can be written in the form*

$$d = am + bn,$$

where m and n are integers.

As in the example worked out above, we can maintain the representation of integers in the form $am + bn$ during the computation. This shows that the expression for d in the theorem not only exists, but is easily computable.

6.7   Congruences

Notation is not part of the bare logical structure of mathematics: we could denote the set of real numbers by $\mathbf{V}$, or addition by #, and the meaning of mathematical results would be the same. But a good notation may be wonderfully suggestive, leading to a real conceptual breakthrough. One such important step was taken when Carl Friedrich Gauss noticed that we use the phrase "a gives the same remainder as b when divided by m" very often, and that this relation behaves quite similarly to equality. He introduced a notation for this, called *congruence*.

FIGURE 6.4. Carl Friedrich Gauss (1777–1855).

If a and b give the same remainder when divided by m (where a, b, m are integers and $m > 0$), then we write

$$a \equiv b \pmod{m}$$

(read: a is congruent to b modulo m). An equivalent way of saying this is that m is a divisor of $b - a$. The number m is called the *modulus* of the congruence relation.

This notation suggests that we want to consider this relation as an analogue of equality. And indeed, many of the properties of equality are valid for congruences, at least if we keep the modulus m fixed. We have *reflexivity*,

$$a \equiv a \pmod{m},$$

symmetry,

$$a \equiv b \pmod{m} \quad \implies \quad b \equiv a \pmod{m},$$

and *transitivity*,

$$a \equiv b \pmod{m}, \qquad b \equiv c \pmod{m} \qquad \Longrightarrow \qquad a \equiv c \pmod{m}.$$

These are trivial if we think of the congruence relation as claiming equality: namely, equality of the remainders when divided by m.

We can compute with congruences just as we can with equations. If we have two congruences with the same modulus,

$$a \equiv b \pmod{m} \quad \text{and} \quad c \equiv d \pmod{m},$$

then we can add them, subtract them, and multiply them to get

$$a + c \equiv b + d \pmod{m}, \quad a - c \equiv b - d \pmod{m}, \quad ac \equiv bd \pmod{m}$$

(we'll return to division later). A useful special case of the multiplication rule is that we can multiply both sides of a congruence by the same number: if $a \equiv b \pmod{m}$, then $ka \equiv kb \pmod{m}$ for every integer k.

These properties need to be proved, however. By hypothesis, $a - b$ and $c - d$ are divisible by m. To see that congruences can be added, we must verify that $(a + c) - (b + d)$ is divisible by m. To this end, we write it in the form $(a - b) + (c - d)$, which shows that it is the sum of two integers divisible by m and so it is also divisible by m.

The proof that congruences can be subtracted is very similar, but multiplication is a bit trickier. We have to show that $ac - bd$ is divisible by m. To this end, we write it in the form

$$ac - bd = (a - b)c + b(c - d).$$

Here $a - b$ and $c - d$ are divisible by m, and hence so are $(a - b)c$ and $b(c - d)$, and hence so is their sum.

The congruence notation is very convenient in formulating various statements and arguments about divisibility. For example, Fermat's Theorem (Theorem 6.5.1) can be stated as follows: If p is a prime then

$$a^p \equiv a \pmod{p}.$$

6.7.1 What is the largest integer m for which $12345 \equiv 54321 \pmod{m}$?

6.7.2 Which of the following "rules" are true?

(a) $a \equiv b \pmod{c} \quad \Rightarrow \quad a + x \equiv b + x \pmod{c + x}$;

(b) $a \equiv b \pmod{c} \quad \Rightarrow \quad ax \equiv bx \pmod{cx}$.

(c) $\left. \begin{array}{l} a \equiv b \pmod{c} \\ x \equiv y \pmod{z} \end{array} \right\} \quad \Rightarrow \quad a + x \equiv b + y \pmod{c + z}$;

(d) $\left. \begin{array}{l} a \equiv b \pmod{c} \\ x \equiv y \pmod{z} \end{array} \right\} \quad \Rightarrow \quad ax \equiv by \pmod{cz}$.

6.7.3 How would you define $a \equiv b \pmod 0$?

6.7.4 (a) Find two integers a and b such that $2a \equiv 2b \pmod 6$, but $a \not\equiv b \pmod 6$. (b) Show that if $c \neq 0$ and $ac \equiv bc \pmod{mc}$, then $a \equiv b \pmod m$.

6.7.5 Let p be a prime. Show that if x, y, u, v are integers such that $x \equiv y \pmod p$, $u, v > 0$, and $u \equiv y \pmod{p-1}$, then $x^u \equiv y^v \pmod p$.

6.8 Strange Numbers

What is Thursday + Friday?

If you don't understand the question, ask a child. He/she will tell you that it is Tuesday. (There may be some discussion as to whether the week starts with Monday or Sunday; but even if we feel it starts with Sunday, we can still say that Sunday is day 0.)

Now we should not have difficulty figuring out that Wednesday · Tuesday = Saturday, Thursday^2 = Tuesday, Monday − Saturday = Tuesday, etc.

This way we can do arithmetic operations with the days of the week: We have introduced a new number system! In this system there are only 7 numbers, which we call Su, Mo, Tu, We, Th, Fr, Sa, and we can carry out addition, subtraction, and multiplication just as with numbers (we could call them Sleepy, Dopey, Happy, Sneezy, Grumpy, Doc, and Bashful; what is important is how the arithmetic operations work).

Not only can we define these operations; they work pretty much like operations with integers. Addition and multiplication are commutative,

$$\text{Tu} + \text{Fr} = \text{Fr} + \text{Tu}, \qquad \text{Tu} \cdot \text{Fr} = \text{Fr} \cdot \text{Tu},$$

and associative,

$$(\text{Mo} + \text{We}) + \text{Fr} = \text{Mo} + (\text{We} + \text{Fr}), \qquad (\text{Mo} \cdot \text{We}) \cdot \text{Fr} = \text{Mo} \cdot (\text{We} \cdot \text{Fr}),$$

and distributive,

$$(\text{Mo} + \text{We}) \cdot \text{Fr} = (\text{Mo} \cdot \text{Fr}) + (\text{We} \cdot \text{Fr}).$$

Subtraction is the inverse of addition:

$$(\text{Mo} + \text{We}) - \text{We} = \text{Mo}.$$

Sunday acts like 0:

$$\text{We} + \text{Su} = \text{We}, \qquad \text{We} \cdot \text{Su} = \text{Su},$$

and Monday acts like 1:

$$\text{We} \cdot \text{Mo} = \text{We}.$$

All this is nothing new if we think of "Monday" as 1, "Tuesday" as 2, etc., and realize that since day 8 is Monday again, we have to replace the result of any arithmetic operation by its remainder modulo 7. All the above identities express congruence relations, and are immediate from the basic properties of congruences.

What about division? In some cases, this is obvious. For example, what is Sa/We? Translating to integers, this is 6/3, which is 2, i.e., Tu. Check: Tu · We = Sa.

But what is Tu/We? In our more familiar number systems, this would be 2/3, which is not an integer; in fact, rational numbers were introduced precisely so that we could talk about the result of all divisions (except divisions by 0). Do we have to introduce "fractional days of the week"?

It turns out that this new number system (with only 7 "numbers") is nicer! What does Tu/We mean? It is a "number" X such that $X \cdot$ We = Tu. But it is easy to check that We · We = Tu; so we have (or at least it seems to make sense to say that we have) that Tu/We = We.

This gives an example showing that we may be able to carry out division without introducing new "numbers" (or new days of the week), but can we always carry out the division? To see how this works, let's take another division: We/Fr, and let's try *not* to guess the result; instead, call it X and show that one of the days of the week must be appropriate for X.

So let $X = $ We/Fr. This means that $X \cdot$ Fr = We. For each day X of the week, the product $X \cdot$ Fr is some day of the week.

The main claim is that *for different days X, the products $X \cdot$ Fr are all different.* Indeed, suppose that

$$X \cdot \text{Fr} = Y \cdot \text{Fr}.$$

Then

$$(X - Y) \cdot \text{Fr} = \text{Su} \tag{6.2}$$

(we used here the distributive law and the fact that Sunday acts like 0). Now, Sunday is analogous to 0 also in the sense that just as the product of two nonzero numbers is nonzero, the product of two non-Sunday days is non-Sunday. (Check!) So we must have $X - Y = $ Su, or $X = Y + $ Su $= Y$.

So the days $X \cdot$ Fr are all different, and there are seven of them, so every day of the week must occur in this form. In particular, "We" will occur.

This argument works for any division, except when we try to divide by Sunday; we already know that Sunday acts like 0, and so Sunday multiplied by any day is Sunday, so we cannot divide any other day by Sunday (and the result of Su/Su is not well defined; it could be any day).

Congruences introduced in Section 6.7 provide an often very convenient way to handle these strange numbers. For example, we can write (6.2) in the form

$$(x - y) \cdot 5 \equiv 0 \pmod 7$$

(where x and y are the numbers corresponding to the days X and Y), and so 7 is a divisor of $(x - y)5$. But 5 is not divisible by 7 and neither is $x - y$ (since these are two different nonnegative integers smaller than 7). Since 7 is a prime, this is a contradiction. This way can talk about ordinary numbers instead of the days of the week; the price we pay is that we have to use congruences instead of equality.

6.8.1 Find We/Fr; Tu/Fr; Mo/Tu; Sa/Tu.

Is there anything special about the number 7 here? In a society where the week consists of 10 or 13 or 365 days, we could define addition, subtraction, and multiplication of the days of the week similarly.

Let m be the number of days of the week, which in mathematical language we call the modulus. It would be impractical to introduce new names for the days of the week,[1] so let's just call them $\bar{0}, \bar{1}, \ldots, \overline{m-1}$. The overlining indicates that, for example, $\bar{2}$ refers not only to day 2, but also to day $m + 2$, day $2m + 2$, etc.

Addition is defined by $\bar{a} + \bar{b} = \bar{c}$, where c is the remainder of $a + b$ modulo m. Multiplication and subtraction are defined in a similar way. This way we have a new number system: It consists of only m numbers, and the basic arithmetic operations can be carried out. These operations will obey the basic laws of computation, which follows just as in the case $m = 7$ above. This version of arithmetic is called *modular arithmetic*.

What about division? If you carefully read the proof that we can do division when $m = 7$, you see that it uses one special property of 7: that it is a prime! There is indeed a substantial difference between modular arithmetic with prime and nonprime moduli.[2] In what follows, we shall restrict our attention to the case where the modulus is a prime, and to emphasize this, we will denote it by p. This number system consisting of $\bar{0}, \bar{1}, \ldots, \overline{p-1}$, with the four operations defined as above, is called a *prime field*.

The 2-element field. The smallest prime number is 2, and the simplest prime field has only 2 elements, $\bar{0}$ and $\bar{1}$. It is easy to give the addition and multiplication tables:

+	$\bar{0}$	$\bar{1}$
$\bar{0}$	$\bar{0}$	$\bar{1}$
$\bar{1}$	$\bar{1}$	$\bar{0}$

$\cdot$	$\bar{0}$	$\bar{1}$
$\bar{0}$	$\bar{0}$	$\bar{0}$
$\bar{1}$	$\bar{0}$	$\bar{1}$

(There is really only one operation here that does not follow from the general properties of 0 and 1, namely, $\bar{1} + \bar{1} = \bar{0}$. There is no need to specify

---

[1] In many languages, the names of some days are derived from numbers.
[2] Plural of "modulus."

the subtraction table, since in this field $a + b = a - b$ for every a and b (check!), nor the division table, since this is obvious: We cannot divide by $\bar{0}$, and dividing by $\bar{1}$ does not change the dividend.)

It is inconvenient to write all these bars over the numbers, so we most often omit them. But then we have to be careful, because we must know whether $1 + 1$ means 2 or 0; therefore, we change the sign of addition, and use $\oplus$ for the addition in the 2-element field. In this notation, the addition and multiplication tables look like this:

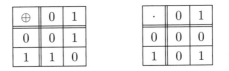

(we did not have to introduce a new multiplication symbol, because the multiplication table for 0 and 1 is the same in the 2-element field as for ordinary numbers).

This field is very small but very important, because a lot of computer science, information theory, and mathematical logic uses it: Its two elements can be interpreted as "YES–NO," "TRUE–FALSE," "SIGNAL–NO SIGNAL," etc.

6.8.2 Let 0 mean "FALSE" and 1 mean "TRUE." Let A and B be two statements (which are either true or false). Express, using the operations $\oplus$ and $\cdot$, the truth of "not A," "A or B," "A and B."

6.8.3 Let the modulus be 6; show by an example that division by a nonzero "number" cannot always be carried out. Generalize the example to every composite modulus.

Division in modular arithmetic. Our argument that division in modular arithmetic can be carried out if the modulus is a prime was reasonably simple but it did not tell us how to carry out the division. To find the quotient by this method would involve looking at all numbers between 0 and $p - 1$, which was OK for $p = 7$, but would be quite tedious for a prime like $p = 234{,}527$ (not to mention the really huge primes used in cryptography and computer security, as we'll see).

So how do we divide, say, $\overline{53}$ by $\bar{2}$ modulo 234,527?

We can simplify the problem, and just ask about dividing $\bar{1}$ by $\bar{2}$ modulo 234,527. If we have that $\bar{1}/\bar{2} = \bar{a}$, then we can get $\overline{53}/\bar{2} = \overline{53} \cdot \bar{a}$, which we know how to compute.

At this point the proof can be explained better in the general case. We are given a prime modulus p and an integer a $(1 \le a \le p - 1)$, and want to find an integer x $(0 \le x \le p - 1)$ such that $\overline{ax} = \bar{1}$. Using the congruence notation from Section 6.7, we can write this as

$$ax \equiv 1 \pmod{p}.$$

The key to solving this problem is the Euclidean Algorithm. Let us compute the greatest common divisor of a and p. This sounds silly, since we know the answer right away: p is a prime and $1 \leq a < p$, so they cannot have any common divisor greater than 1, and so $\gcd(p, a) = 1$. But recall that the Euclidean Algorithm gives more: it will provide the greatest common divisor in the form $au + pv$, where u ad v are integers. Thus we get

$$au + pv = 1,$$

which implies that

$$au \equiv 1 \pmod{p}.$$

We are almost done; the only problem is that the integer u may not be between 1 and $p - 1$. But if x is the remainder of u modulo p, then multiplying the congruence $x \equiv u \pmod{p}$ by a (recall from Section 6.7 that this is a legal operation on congruences), we get

$$ax \equiv au \equiv 1 \pmod{p},$$

and since $0 \leq x \leq p - 1$, this solves our problem.

Let us follow this algorithm on our example above, with $a = 2$ and $p = 234{,}527$. The Euclidean Algorithm works really simply in this case: Divide 234,527 by 2 with remainder, and the remainder is already down to 1. This gives

$$2 \cdot (-117{,}263) + 234{,}527 \cdot 1 = 1.$$

The remainder of -117,263 modulo 234,527 is 117,264, so we get that

$$\overline{1/2} = \overline{117{,}264}.$$

6.8.4 Compute $\overline{1/53}$ modulo 234527.

Once we know how to do basic arithmetic, more involved tasks like solving linear equations can be done by recalling what we would do with ordinary numbers. We illustrate this by some examples, where we use the congruence notation along with its basic properties from Section 6.7.

Example 1. Consider a linear equation, say

$$\overline{7}X + \overline{3} = \overline{0},$$

where the modulus is 47 (check in the table that this is a prime!). We can rewrite this as a congruence:

$$7x + 3 \equiv 0 \pmod{47}.$$

This second form is the more usual, so let's work with this.

Just as we would do with an equation, we transform this as

$$7x \equiv -3 \pmod{47} \tag{6.3}$$

(we could replace -3 by its remainder 44 modulo 47 if we wanted to keep numbers positive, but this is optional).

Next we have to find the reciprocal of 7 modulo 47. The Euclidean Algorithm gives

$$\gcd(7, 47) = \gcd(7, 5) = \gcd(2, 5) = \gcd(2, 1) = 1,$$

and following the extended version we get

$$5 = 47 - 6 \cdot 7, \qquad 2 = 7 - 5 = 7 - (47 - 6 \cdot 7) = 7 \cdot 7 - 47,$$

$$1 = 5 - 2 \cdot 2 = (47 - 6 \cdot 7) - 2 \cdot (7 \cdot 7 - 47) = 3 \cdot 47 - 20 \cdot 7,$$

which shows that $(-20) \cdot 7 \equiv 1 \pmod{47}$. So the reciprocal of 7 modulo 47 is -20 (which again we could write as 27).

Now dividing both sides of (6.3) by 7, which is the same as multiplying both sides by 27, we get

$$x \equiv 13 \pmod{47}.$$

(Here we get 13 either as the remainder of $(-3)(-20)$, or as the remainder of $44 \cdot 27$ modulo 47; the result is the same.)

Example 2. Next, let us solve a system of two linear equations, with two variables. We'll make the numbers a little bigger, to see that we can cope with larger numbers too. Let the modulus be $p = 127$, and consider the equations

$$\overline{12}X + \overline{31}Y = \overline{2}, \qquad (6.4)$$
$$\overline{2}X + \overline{89}Y = \overline{23}.$$

We can rewrite these as congruences:

$$12x + 31y \equiv 2 \pmod{127},$$
$$2x + 89y \equiv 23 \pmod{127}.$$

a. Eliminate a variable. How would we solve this system if these were ordinary equations? We could multiply the second equation by 6 and subtract it from the first, to eliminate the x terms. We can do this in this prime field as well, and get

$$(31 - 6 \cdot 89)y \equiv 2 - 6 \cdot 23 \pmod{127},$$

or

$$(-503)y \equiv -136 \pmod{127}.$$

We can replace these negative numbers by their remainders modulo 127 to get

$$5y \equiv 118 \pmod{127}. \qquad (6.5)$$

Division. Next, we want to divide the equation by 5. This is what we discussed above: We have to use the Euclidean Algorithm. The computation of the greatest common divisor is easy:

$$\gcd(127, 5) = \gcd(2, 5) = \gcd(2, 1) = 1.$$

This does not give anything new: We knew in advance that this greatest common divisor will be 1. To get more, we have to follow this computation by another one, where each number is written as an integer multiple of 127 plus an integer multiple of 5:

$$\gcd(127, 5) = \gcd(127 - 25 \cdot 5, 5) = \gcd(127 - 25 \cdot 5, (-2) \cdot 127 + 51 \cdot 5) = 1.$$

This gives that

$$(-2) \cdot 127 + 51 \cdot 5 = 1.$$

Thus $5 \cdot 51 \equiv 1 \pmod{127}$, and so we have found the "reciprocal" of 5 modulo 127.

Instead of dividing equation (6.4) by five, we multiply by its "reciprocal," 51, to get

$$y \equiv 51 \cdot 118 \pmod{127}. \tag{6.6}$$

Conclusion. If we evaluate the right-hand side of (6.6) and then compute its remainder modulo 127, we get that $y \equiv 49 \pmod{127}$, or in other words, $Y = \overline{49}$ is the solution. To get x, we have to substitute this value back into one of the original equations:

$$2x + 89 \cdot 49 \equiv 23 \pmod{127},$$

whence

$$2x \equiv 23 - 89 \cdot 49 \equiv 107 \pmod{127}.$$

So we have to do one more division. In analogy with what we did above, we get

$$(-63) \cdot 2 + 127 = 1,$$

and hence

$$64 \cdot 2 \equiv 1 \pmod{127}.$$

So instead of dividing by 2, we can multiply by 64, to get

$$x \equiv 64 \cdot 107 \pmod{127}.$$

Computing the right-hand side and its remainder modulo 127, we get that $x \equiv 117 \pmod{127}$, or in other words, $X = \overline{117}$. Thus we have solved (6.4).

Example 3. We can even solve some quadratic equations; for example,

$$x^2 - 3x + 2 \equiv 0 \pmod{53}.$$

We can write this as

$$(x - 1)(x - 2) \equiv 0 \pmod{53}.$$

One of the factors on the left-hand side must be congruent to 0 modulo 53, whence either $x \equiv 1 \pmod{53}$ or $x \equiv 2 \pmod{53}$.

Here we found a way to write the left-hand side as a product just by looking at it. What happens if we have an equation with larger numbers, say $x^2 + 134517x + 105536 \equiv 0 \pmod{234527}$? We doubt that anybody can guess a decomposition. In this case, we can try to follow the high-school procedure for solving quadratic equations. This works, but one step of it is quite difficult: taking square roots. This can be done efficiently, but the algorithm is too complicated to be included here.

6.8.5 Solve the congruence system

$$2x + 3y \equiv \qquad 1 \pmod{11},$$
$$x + 4y \equiv \qquad 4 \pmod{11}.$$

6.8.6 Solve the "congruence equations"

(a) $x^2 - 2x \equiv 0 \pmod{11}$,   (b) $x^2 \equiv 4 \pmod{23}$.

6.9 Number Theory and Combinatorics

Many of the combinatorial tools that we have introduced are very useful in number theory as well. Induction is used all over the place. We show some elegant arguments based on the *Pigeonhole Principle* and on *inclusion–exclusion*.

> *We are given n natural numbers: $a_1, a_2, \ldots, a_n$. Show that we can choose a (nonempty) subset of these numbers whose sum is divisible by n.*

(It is possible that this subset contains all n numbers.)

Solution. Consider the following n numbers:

$$b_1 = a_1,$$
$$b_2 = a_1 + a_2,$$
$$b_3 = a_1 + a_2 + a_3,$$
$$\vdots$$
$$b_n = a_1 + a_2 + a_3 + \cdots + a_n.$$

If there is a number among these n numbers that is divisible by n, then we have found what we want. If there is none, then let us divide all the numbers $b_1, b_2, \ldots, b_n$ by n with residue. Write down these residues. What are the numbers we were getting? It could be $1, 2, \ldots,$ or $n-1$. But we have a total of n numbers! So by the pigeonhole principle, there will be two numbers among $b_1, b_2, \ldots, b_n$ that give the same residue when we divide them by n. Say these two numbers are b_i and b_j $(i < j)$. Then their difference $b_j - b_i$ is divisible by n. But

$$b_j - b_i = a_{i+1} + a_{i+2} + \cdots + a_j.$$

So we have found a special subset of the numbers $a_1, a_2, \ldots, a_n$, namely $a_{i+1}, a_{i+2}, \ldots, a_j$, whose sum is divisible by n. And this is what we wanted to prove.

6.9.1 We are given n numbers from the set $\{1, 2, \ldots, 2n - 1\}$. Prove that we can always find two numbers among these n numbers that are relatively prime to each other.

As a very important application of inclusion–exclusion, let's answer the following question about numbers: *How many numbers are there up to* 1200 *that are relatively prime to* 1200*?*

Since we know the prime factorization of 1200,namely, $1200 = 2^4 \cdot 3 \cdot 5^2$, we know that the numbers divisible by any of 2, 3, or 5 are precisely those that have a common divisor with 1200. So we are interested in counting the positive integers smaller than 1200 and not divisible by 2, 3, or 5.

One can easily compute that up to 1200, there are

$$\frac{1200}{2}$$ numbers divisible by 2

(every second number is even),

$$\frac{1200}{3}$$ numbers divisible by 3,

$$\frac{1200}{5}$$ numbers divisible by 5.

Those numbers divisible by both 2 and 3 are just those that are divisible by 6. Therefore up to 1200 there are

$$\frac{1200}{6}$$ numbers divisible by 2 and 3,

and similarly, there are

$$\frac{1200}{10}$$ numbers divisible by 2 and 5,

$$\frac{1200}{15}$$ numbers divisible by 3 and 5.

Finally, the numbers divisible by all of 2, 3, 5 are precisely those that are divisible by 30; so there are

$$\frac{1200}{30} \text{ numbers divisible by all of 2, 3, 5.}$$

Now with these data, we can use inclusion–exclusion to compute the number we are looking for:

$$1200 - \left(\frac{1200}{2} + \frac{1200}{3} + \frac{1200}{5}\right) + \frac{1200}{2 \cdot 3} + \frac{1200}{2 \cdot 5} + \frac{1200}{3 \cdot 5} - \frac{1200}{2 \cdot 3 \cdot 5} = 320.$$

If we pull out 1200 from the left-hand side of the above equality, what remains can be transformed into a nice product form (check the calculations!):

$$1200 \cdot \left(1 - \frac{1}{2} - \frac{1}{3} - \frac{1}{5} + \frac{1}{2 \cdot 3} + \frac{1}{2 \cdot 5} + \frac{1}{3 \cdot 5} - \frac{1}{2 \cdot 3 \cdot 5}\right)$$

$$= 1200 \cdot \left(1 - \frac{1}{2}\right) \cdot \left(1 - \frac{1}{3}\right) \cdot \left(1 - \frac{1}{5}\right).$$

Let n be a natural number. We denote by $\phi(n)$ the number of those numbers that are not larger than n and are relatively prime to n (we used here "not larger," instead of "smaller," which has significance only if $n = 1$, since this is the only case when the number itself is relative prime to itself; so $\phi(1) = 1$). Primes, of course, have the most numbers relatively prime to them: If p is a prime, then every smaller positive integer is counted in $\phi(p)$, so $\phi(p) = p - 1$. In general, the number $\phi(n)$ can be computed as we did in the concrete case above: *if $p_1, p_2, \ldots, p_r$ are the different prime factors of n, then*

$$\phi(n) = n \cdot \left(1 - \frac{1}{p_1}\right) \cdot \left(1 - \frac{1}{p_2}\right) \cdots \left(1 - \frac{1}{p_r}\right). \tag{6.7}$$

The proof follows the calculations above, and is given as Exercise 6.9.2.

6.9.2 Prove (6.7).

6.9.3 Let n be a natural number. We compute $\phi(d)$ of every divisor d of n, then add up all these numbers. What is the sum? (Experiment, formulate a conjecture, and prove it.)

6.9.4 We add up all the positive integers smaller than n and relatively prime to n. What do we get?

6.9.5 Prove the following extension of Fermat's Theorem: If $\gcd(a, b) = 1$, then $a^{\phi(b)} - 1$ is divisible by b.
    [Hint: Generalize the proof of Fermat's Theorem in Exercise 6.5.4.]

6.10   How to Test Whether a Number is a Prime?

Is 123,456 a prime? Of course not, it is even. Is 1,234,567 a prime? This is not so easy to answer, but if you are hard pressed, you can try all numbers $2, 3, 4, 5 \ldots$ to see if they are divisors. If you have the patience to go up to 127, then you are done: $1234567 = 127 \cdot 9721$.

What about 1,234,577? Again you can try to find a divisor by trying out $2, 3, 4, 5, \ldots$. But this time you don't find a proper divisor! Still, if you are really patient and keep on until you get to the square root of 1234577, which is $1111.1 \ldots$, you know that you are not going to find a proper divisor (why?).

Now what about the number

$$1,111,222,233,334,444,555,566,667,777,888,899,967?$$

If this is a prime (as it is), then we have to to try out all numbers up to its square root; since the number is larger than 10^{36}, its square root is larger than 10^{18}. Trying out more than 10^{18} numbers is a hopeless task even for the world's most powerful computer.

The Fermat test. So how do we know that this number is a prime? Well, our computer tells us, but how does the computer know? An approach is offered by Fermat's Theorem. Its simplest nontrivial case says that *if p is a prime, then $p \mid 2^p - 2$.* If we assume that p is odd (which only excludes the case $p = 2$), then we also know that $p \mid 2^{p-1} - 1$.

What happens if we check the divisibility relation $n \mid 2^{n-1} - 1$ for composite numbers? It obviously fails if n is even (no even number is a divisor of an odd number), so let's restrict our attention to odd numbers. Here are some results:

$$9 \nmid 2^8 - 1 = 255, \qquad 15 \nmid 2^{14} - 1 = 16{,}383, \qquad 21 \nmid 2^{20} - 1 = 1{,}048{,}575,$$

$$25 \nmid 2^{24} - 1 = 16{,}777{,}215.$$

This suggests that perhaps we could test whether the number n is a prime by checking whether the relation $n \mid 2^{n-1} - 1$ holds. This is a nice idea, but it has several major shortcomings.

How to compute LARGE powers. It is easy to write up the formula $2^{n-1} - 1$, but it is quite a different matter to compute it! It seems that to get 2^{n-1}, we have to multiply by 2 $n - 2$ times. For a 100-digit number n, this is about 10^{100} steps, which we will never be able to carry out.

But we can be tricky when we compute 2^{n-1}. Let us illustrate this on the example of 2^{24}: We could start with $2^3 = 8$, square it to get $2^6 = 62$, square it again to get $2^{12} = 4096$, and square it once more to get $2^{24} = 16{,}777{,}216$. Instead of 23 multiplications, we needed only 5.

It seems that this trick worked only because 24 was divisible by such a large power of 2, and so we could compute 2^{24} by repeated squaring, starting from a small number. Let us show how to do a similar trick if the exponent is a less friendly integer, say 29. Here is a way compute 2^{29}:

$$2^2 = 4, \quad 2^3 = 8, \quad 2^6 = 64, \quad 2^7 = 128, \quad 2^{14} = 16{,}384,$$

$$2^{28} = 268{,}435{,}456, \quad 2^{29} = 536{,}870{,}912.$$

It is perhaps best to read this sequence backwards: If we have to compute an odd power of 2, we obtain it by multiplying the previous power by 2; if we have to compute an even power, we obtain it by squaring the appropriate smaller power.

6.10.1 Show that if n has k bits in base 2, then 2^n can be computed using fewer than $2k$ multiplications.

How to avoid LARGE numbers. We have shown how to overcome the first difficulty; but the computations above reveal the second: the numbers grow too large! Let's say that n has 100 digits; then not only is 2^{n-1} astronomical, the number of its digits is astronomical! We could never write it down, let alone check whether it is divisible by n.

The way out is to divide by n as soon as we get any number that is larger than n, and just work with the remainder of the division (or we could say we work in modular arithmetic with modulus n; we won't have to do divisions, so n does not have to be a prime). For example, if we want to check whether $25 \mid 2^{24} - 1$, then we have to compute 2^{24}. As above, we start with computing $2^3 = 8$, then square it to get $2^6 = 64$. We immediately replace it by the remainder of the division $64 \div 25$, which is 14. Then we compute 2^{12} by squaring 2^6, but instead we square 14 to get 196, which we replace by the remainder of the division $196 \div 25$, which is 21. Finally, we obtain 2^{24} by squaring 2^{12}, but instead we square 21 to get 441, and then divide this by 25 to get the remainder 16. Since $16 - 1 = 15$ is not divisible by 25, it follows that 25 is not a prime.

This does not sound like an impressive conclusion, considering the triviality of the result, but this was only an illustration. If n has k bits in base 2, then as we have seen, it takes only $2k$ multiplications to compute 2^n, and all we have to do is one division (with remainder) in each step to keep the numbers small. We never have to deal with numbers larger than n^2. If n has 100 digits, then n^2 has 199 or 200; not much fun to multiply by hand, but quite easily manageable by computers.

Pseudoprimes. But here comes the third shortcoming of the primality test based on Fermat's Theorem. Suppose that we carry out the test for a number n. If it fails (that is, n is not a divisor of $2^{n-1} - 1$), then of course we know that n is not a prime. But suppose we find that $n \mid 2^{n-1} - 1$. Can we conclude that n is a prime? Fermat's Theorem certainly does not justify this conclusion. Are there composite numbers n for which $n \mid 2^{n-1} - 1$?

Unfortunately, the answer is yes. The smallest such number is $341 = 11 \cdot 31$. This is not a prime, but it satisfies

$$341 \mid 2^{340} - 1. \tag{6.8}$$

(How do we know that this divisibility relation holds without extensive computation? We can use Fermat's Theorem. It is sufficient to argue that both 11 and 31 are divisors of $2^{340} - 1$, since then so is their product, 11 and 31 being different primes. By Fermat's Theorem,

$$11 \mid 2^{10} - 1.$$

Next we invoke the result of Exercise 6.1.6: It implies that

$$2^{10} - 1 \mid 2^{340} - 1.$$

Hence

$$11 \mid 2^{340} - 1.$$

For 31, we don't need Fermat's Theorem, but only exercise (6.1.6) again:

$$31 = 2^5 - 1 \mid 2^{340} - 1.$$

This proves (6.8).)

Such numbers, which are not primes but behave like primes in the sense that Fermat's Theorem with base $a = 3$ holds true for them, are called *pseudoprimes* (fake primes), or more precisely, pseudoprimes to base 2. While such numbers are quite rare (there are only 22 pseudoprimes to base 2 between 1 and 10,000), they do show that our primality test can give a "false positive," and thus (in a strict mathematical sense) it is not a primality test at all.

(If we can afford to make an error every now and then, then we can live with the simple Fermat test with base 2. If the worst that can happen when a composite number is believed to be a prime is that a computer game crashes, we can risk this; if the security of a bank, or a country, depends on not using a fake prime, we have to find something better.)

One idea that comes to the rescue is that we haven't used the full force of Fermat's Theorem: We can also check that $n \mid 3^n - 3$, $n \mid 5^n - 5$, etc. These tests can be carried out using the same tricks as described above. And in fact, already the first of these tests rules out the "fake prime" 341: it is not a divisor of $3^{340} - 1$.

The following observation tells us that this always works, at least if we are patient enough:

> *A positive integer $n > 1$ is a prime if and only if it passes the Fermat test*
>
> $$n \mid a^{n-1} - 1$$
>
> *for every base $a = 1, 2, 3, \ldots, n - 1$.*

Fermat's Theorem tells us that primes do pass the Fermat test for every base. On the other hand, if n is composite, then there are numbers a, $1 \leq a \leq n-1$, that are not relatively prime to n, and every such a will fail the Fermat test: Indeed, if p is a common prime divisor of a and n, then p is a divisor of a^{n-1}, so it cannot be a divisor of $a^{n-1} - 1$, and hence n cannot be a divisor of $a^{n-1} - 1$.

But this general Fermat test is not efficient enough. Imagine that we are given a natural number n, with a few hundred digits, and we want to test whether or not it is a prime. We can carry out the Fermat test with base 2. Suppose it passes. Then we can try base 3. Suppose it passes again, etc. How long do we have to go before we can conclude that n is a prime? Looking at the argument above justifying the general Fermat test, we see that we don't have to go farther than the first number having a common divisor with n. It is easy to see that the smallest such number is the least prime divisor of n. For example, if $n = pq$, where p and q are distinct primes, having say 100 digits each (so n has 199 or 200 digits), then we have to try everything up to the smaller of p and q, which is more than 10^{99} trials, which is hopelessly large. (And furthermore, if we go this far anyway, we may do a simple divisibility test, no need for anything fancy like Fermat's Theorem!)

Instead of starting with 2, we could start checking whether Fermat's Theorem holds with any other base a; for example, we could choose a random integer a in the range $1 \leq a \leq n-1$. We know that it fails if we hit any a that is not relatively prime to n. Does this give us a good chance of discovering that n is not a prime if in fact it is not? This depends on n, but certain values of n are definitely bad. For example, suppose that $n = pq$ where p and q are different primes. It is easy to list those numbers a that are not relatively prime to n: these are the multiples of p ($p, 2p, \ldots, (q-1)p, qp$) and the multiples of q ($q, 2q, \ldots, (p-1)q, pq$). The total number of such numbers a is $q + p - 1$ (since $pq = n$ occurs on both lists). This number is larger than $2 \cdot 10^{99}$, but less than $2 \cdot 10^{100}$, and so the probability that we hit one of these number when we choose a random a is less than

$$\frac{2 \cdot 10^{100}}{10^{199}} = 2 \cdot 10^{-99},$$

which shows that this event has way too small a probability to ever happen in practice.

Carmichael numbers. Our next hope is that perhaps for a composite number n, the Fermat test will fail much earlier than its smallest prime divisor, or else for a random choice of a it will fail for many other numbers besides those not relatively prime to n. Unfortunately, this is not always so. There are integers n, called *Carmichael numbers*, which are even worse than pseudoprimes: They pass the Fermat test for every base a relatively prime to n. In other words, they satisfy

$$n \mid a^{n-1} - 1$$

for every a such that $\gcd(n, a) = 1$. The smallest such number is $n = 561$. While such numbers are very rare, they do show that the Fermat test is not completely satisfactory.

The Miller–Rabin test. But in the late 1970's, M. Rabin and G. Miller found a very simple way to strengthen Fermat Theorem just a little bit, and thereby overcome the difficulty caused by Carmichael numbers. We illustrate the method on the example of 561. We use some high-school math, namely, the identity $x^2 - 1 = (x - 1)(x + 1)$, to factor the number $a^{560} - 1$:

$$
\begin{aligned}
a^{560} - 1 &= \left(a^{280} - 1\right)\left(a^{280} + 1\right) \\
&= \left(a^{140} - 1\right)\left(a^{140} + 1\right)\left(a^{280} + 1\right) \\
&= \left(a^{70} - 1\right)\left(a^{70} + 1\right)\left(a^{140} + 1\right)\left(a^{280} + 1\right) \\
&= \left(a^{35} - 1\right)\left(a^{35} + 1\right)\left(a^{70} + 1\right)\left(a^{140} + 1\right)\left(a^{280} + 1\right).
\end{aligned}
$$

Now suppose that 561 were a prime. Then by Fermat's "Little" Theorem, it would have to divide $a^{560} - 1$ for every $1 \le a \le 560$. If a prime divides a product, it divides one of the factors (exercise 6.3.3), and hence at least one of the relations

$$561 \mid a^{35} - 1 \quad 561 \mid a^{35} + 1 \quad 561 \mid a^{70} + 1 \quad 561 \mid a^{140} + 1 \quad 561 \mid a^{280} + 1$$

must hold. But already for $a = 2$, none of these relations hold.

The Miller–Rabin test is an elaboration of this idea. Given an odd integer $n > 1$ that we want to test for primality, we choose an integer a from the range $0 \le a \le n - 1$ at random, and consider $a^n - a$. We factor it as $a(a^{n-1} - 1)$, and then go on to factor it, using the identity $x^2 - 1 = (x - 1)(x + 1)$, as long as we can. Then we test that one of the factors must be divisible by n.

If the test fails, we can be sure that n is not a prime. But what happens if it succeeds? Unfortunately, this can still happen even if n is composite; but the crucial point is that *this test gives a false positive with probability less than* $\frac{1}{2}$ (remember that we chose a random a).

Reaching a wrong conclusion half of the time does not sound so good at all; but we can repeat the experiment several times. If we repeat it 10 times (with a new randomly chosen a each time), the probability of a false positive is less than $2^{-10} < 1/1000$ (since to conclude that n is prime, all 10 runs must give a false positive, independently of each other). If we repeat the experiment 100 times, the probability of a false positive drops below $2^{-100} < 10^{-30}$, which is astronomically small.

So this algorithm, when repeated sufficiently often, tests primality with error probability that is much less than the probability of, say, hardware failure, and therefore it is quite adequate for practical purposes. It is widely used in programs like Maple and Mathematica and in cryptography.

Suppose that we test a number n for primality and find that it is composite. Then we would like to find its prime factorization. It is easy to see that instead of this, we could ask for less: for a decomposition of n into the product of two smaller positive integers, $n = ab$. If we have a method to find such a decomposition efficiently, then we can go on and test a and b for primality. If they are primes, we have found the prime factorization of n; if (say) a is not a prime, we can use our method to find a decomposition of a into the product of two smaller integers, etc. Since n has at most $\log_2 n$ prime factors (Exercise 6.3.4), we have to repeat this at most $\log_2 n$ times (which is less than the number of its bits).

But unfortunately (or fortunately? see Chapter 15 on cryptography), no efficient method is known to write a composite number as a product of two smaller integers. It would be very important to find an efficient factorization method, or to give a mathematical proof that no such method exists; but we don't know what the answer is.

6.10.2 Show that 561 is a Carmicheal number; more exactly, show that $561 \mid a^{561} - a$ for every integer a. [Hint: Since $561 = 3 \cdot 11 \cdot 17$, it suffices to prove that $3 \mid a^{561} - a$, $11 \mid a^{561} - 1$ and $17 \mid a^{561} - a$. Prove these relations separately, using the method of the proof of the fact that $341 \mid 2^{340} - 1$.]

Review Exercises

6.10.3 Prove that if $c \neq 0$ and $ac \mid bc$, then $a \mid b$.

6.10.4 Prove that if $a \mid b$ and $a \mid c$, then $a \mid b^2 + 3c + 2^b c$.

6.10.5 Prove that every prime larger than 3 gives a remainder of 1 or -1 if divided by 6.

6.10.6 Let $a > 1$, and $k, n > 0$. Prove that $a^k - 1 \mid a^n - 1$ if and only if $k \mid n$.

6.10.7 Prove that if $a > 3$, then a, $a + 2$, and $a + 4$ cannot be all primes. Can they all be powers of primes?

6.10.8 How many integers are there that are not divisible by any prime larger than 20 and not divisible by the square of any prime?

6.10.9 Find the prime factorization of (a) $\binom{20}{10}$; (b) 20!.

6.10.10 Show that a number with 30 digits cannot have more than 100 prime factors.

6.10.11 Show that a number with 160 digits has a prime power divisor that is at least 100. This is not true if we want a prime divisor that is at least 100.

6.10.12 Find the number of (positive) divisors of n, for $1 \leq n \leq 20$ (example: 6 has 4 divisors: 1, 2, 3, 6). Which of these numbers have an odd number of divisors? Formulate a conjecture and prove it.

6.10.13 Find the g.c.d. of 100 and 254, using the Euclidean Algorithm.

6.10.14 Find pairs of integers for which the Euclidean Algorithm lasts (a) 2 steps; (b) 6 steps.

6.10.15 Recalling the Lucas numbers L_n introduced in Exercise 4.3.2, prove the following:

(a) $\gcd(F_{3k}, L_{3k}) = 2$;

(b) if n is not a multiple of 3, then $\gcd(F_n, L_n) = 1$;

(c) $L_{6k} \equiv 2 \pmod 4$.

6.10.16 Prove that for every positive integer m there is a Fibonacci number divisible by m (well, of course, $F_0 = 0$ is divisible by any m; we mean a larger one).

6.10.17 Find integers x and y such that $25x + 41y = 1$.

6.10.18 Find integers x and y such that

$$2x + y \equiv 4 \pmod{17},$$

$$5x - 5y \equiv 9 \pmod{17}.$$

6.10.19 Prove that $\sqrt[3]{5}$ is irrational.

6.10.20 Prove that the two forms of Fermat's Theorem, Theorem 6.5.1 and (6.1), are equivalent.

6.10.21 Show that if $p > 2$ is a prime modulus, then

$$\frac{\overline{1}}{\overline{2}} = \frac{\overline{p+1}}{2}.$$

6.10.22 We are given $n + 1$ numbers from the set $\{1, 2, \ldots, 2n\}$. Prove that there are two numbers among them such that one divides the other.

6.10.23 What is the number of positive integers not larger than 210 and not divisible by 2, 3 or 7?

7
Graphs

7.1 Even and Odd Degrees

We start with the following exercise (admittedly of no practical significance).

> Prove that at a party with 51 people, there is always a person
> who knows an even number of others.

(We assume that acquaintance is mutual. There may be people who don't know each other. There may even be people who don't know anybody else. Of course, such people know an even number of others, so the assertion is true if there is such a person.)

If you don't have any idea how to begin a solution, you should try to experiment. But how to experiment with such a problem? Should we find 51 names for the participants, then create, for each person, a list of those people he or she knows? This would be very tedious, and we would be lost among the data. It would be good to experiment with smaller numbers. But which number can we take instead of 51? It is easy to see that 50, for example, would not do: If, say, we have 50 people who all know each other, then everybody knows 49 others, so there is no person with an even number of acquaintances. For the same reason, we could not replace 51 by 48, or 30, or any even number. Let's hope that this is all; let's try to prove that

> at a party with an odd number of people, there is always a person
> who knows an even number of others.

Now we can at least experiment with smaller numbers. Let us have, say, 5 people: Alice, Bob, Carl, Diane, and Eve. When they first met, Alice knew everybody else; Bob and Carl knew each other, and Carl also knew Eve. So the numbers of acquaintances are: Alice 4, Bob 2, Carl 3, Diane 1, and Eve 2. We have not only one but three people with an even number of acquaintances.

It is still rather tedious to consider examples by listing people and listing pairs knowing each other, and it is quite easy to make mistakes. We can, however, find a graphic illustration that helps a lot. We represent each person by a point in the plane (well, by a small circle, to make the picture nicer), and we connect two of these points by a segment if the people know each other. This simple drawing contains all the information we need (Figure 7.1).

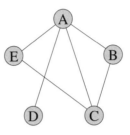

FIGURE 7.1. The graph depicting acquaintance between our friends

A picture of this kind is called a *graph*. More exactly, a graph consists of a set of *nodes* (also known as *points*, or *vertices*) with some pairs of these (not necessarily all pairs) connected by *edges*. It does not matter whether these edges are straight or curvy; all that is important is which pairs of nodes they connect. The set of nodes of a graph G is usually denoted by V; the set of edges, by E. Thus we write $G = (V, E)$ to indicate that the graph G has node set V and edge set E.

The only thing that matters about an edge is the pair of nodes it connects; hence the edges can be considered as 2-element subsets of V. This means that the edge connecting nodes u and v is just the set $\{u, v\}$. We'll further simplify notation and denote this edge by uv.

Can two edges connect the same pair of nodes (parallel edges)? Can an edge connect a node to itself (loop)? The answer to these questions is, of course, our decision. In some applications it is advantageous to allow such edges; in others, they must be excluded. In this book, we generally assume that a pair of nodes is connected by at most one edge, and no node is connected to itself. Such graphs are often called *simple graphs*. If parallel edges are allowed, the graph is often called a *multigraph* to emphasize this fact.

If two nodes are connected by an edge, then they are called *adjacent*. Nodes adjacent to a given node v are called its *neighbors*.

Coming back to our problem, we see that we can represent the party by a graph very conveniently. Our concern is the number of people known by a given person. We can read this off the graph by counting the number of edges leaving a given node. This number is called the *degree* of the node. The degree of node v is denoted by $d(v)$. So A has degree 4, B has degree 2, etc. If Frank now arrives, and he does not know anybody, then we add a new node that is not connected to any other node. So this new node has degree 0.

In the language of graph theory, we want to prove the following:

> *If a graph has an odd number of nodes, then it has a node with even degree.*

Since it is much easier to experiment with graphs than with tables of acquaintances, we can draw many graphs with an odd number of nodes, and count the number of nodes with even degree (Figure 7.2). We find that they contain $5, 1, 1, 7, 3, 3$ such nodes (the last one is a single graph on 7 nodes, not two graphs). So we observe that not only is there always such a node, but the number of such nodes is odd.

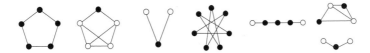

FIGURE 7.2. Some graphs with an odd number of nodes. Black circles mark nodes of even degree.

Now, this is a case in which it is easier to prove more: If we formulate the following stronger statement,

> *If a graph has an odd number of nodes, then the number of nodes with even degree is odd,*

then we made an important step towards the solution! (Why is this statement stronger? Because 0 is not an odd number.) Let's try to find an even stronger statement by looking also at graphs with an even number of nodes. Experimenting on several small graphs again (Figure 7.3), we find that the number of nodes with even degree is $2, 4, 0, 6, 2, 4$. So we conjecture the following:

> *if a graph has an even number of nodes, then the number of nodes with even degree is even.*

This is nicely parallel to the statement about graphs with an odd number of nodes, but it would be better to have a single common statement for the odd and even case. We get such a version if we look at the number of nodes with *odd*, rather than *even*, degree. This number is obtained by subtracting

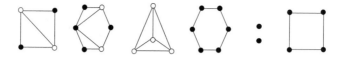

FIGURE 7.3. Some graphs with an even number of nodes. Black circles mark nodes of even degree.

the number of nodes with even degree from the total number of nodes, and hence both statements will be implied by the following:

Theorem 7.1.1 *In every graph, the number of nodes with odd degree is even.*

So what we have to prove is this theorem. It seems that having made the statement stronger and more general in several steps, we have made our task harder and harder. But in fact, we have gotten closer to the solution.

Proof. One way of proving the theorem is to build up the graph one edge at a time, and observe how the parities of the degrees change. An example is shown in Figure 7.4. We start with a graph with no edge, in which every degree is 0, and so the number of nodes with odd degree is 0, which is an even number.

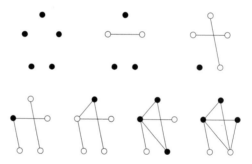

FIGURE 7.4. Building up a graph one edge at a time. Black circles mark nodes of even degree.

Now if we connect two nodes by a new edge, we change the parity of the degrees at these nodes. In particular,

— if both endpoints of the new edge had even degree, we increase the number of nodes with odd degree by 2;

— if both endpoints of the new edge had odd degree, we decrease the number of nodes with odd degree by 2;

— if one endpoint of the new edge had even degree and the other had odd degree, then we don't change the number of nodes with odd degree.

Thus if the number of nodes with odd degree was even before adding the new edge, it remained even after this step. This proves the theorem. (Note that this is a proof by induction on the number of edges.)   □

Graphs are very handy in representing a large variety of situations, not only parties. It is quite natural to consider the graph whose nodes are towns and whose edges are highways (or railroads, or telephone lines) between these towns. We can use a graph to describe an electrical network, say the printed circuit on a card in your computer.

In fact, graphs can be used in any situation where a "relation" between certain objects is defined. Graphs are used to describe bonds between atoms in a molecule, connections between cells in the brain, descent between species, etc. Sometimes the nodes represent more abstract things: For example, they may represent stages of a large construction project, and an edge between two stages means that one arises from the other in a single phase of work. Or the nodes can represent all possible positions in a game (say, chess, although you don't really want to draw this graph), where we connect two nodes by an edge if one can be obtained from the other in a single move.

7.1.1 Find all graphs with $2, 3$, and 4 nodes.

7.1.2   (a) Is there a graph on 6 nodes with degrees $2, 3, 3, 3, 3, 3$?

(b) Is there a graph on 6 nodes with degrees $0, 1, 2, 3, 4, 5$?

(c) How many graphs are there on 4 nodes with degrees $1, 1, 2, 2$?

(d) How many graphs are there on 10 nodes with degrees $1, 1, 1, 1, 1, 1, 1, 1, 1, 1$?

7.1.3 At the end of the party with n people, everybody knows everybody else. Draw the graph representing this situation. How many edges does it have?

7.1.4   (a) Draw a graph with nodes representing the numbers $1, 2, \ldots, 10$, in which two nodes are connected by an edge if and only if one is a divisor of the other.

(b) Draw a graph with nodes representing the numbers $1, 2, \ldots, 10$, in which two nodes are connected by an edge if and only if they have no common divisor larger than 1.[1]

(c) Find the number of edges and the degrees in these graphs, and check that Theorem 7.1.1 holds.

7.1.5 What is the largest number of edges a graph with 10 nodes can have?

---

[1] This is an example where *loops* could play a role: Since $\gcd(1, 1) = 1$ but $\gcd(k, k) > 1$ for $k > 1$, we could connect 1 to itself by a loop, if we allowed loops at all.

7.1.6 How many graphs are there on 20 nodes? (To make this question precise, we have to make sure we know what it means that two graphs are the same. For the purpose of this exercise, we consider the nodes given, and labeled, say, as Alice, Bob, ... The graph consisting of a single edge connecting Alice and Bob is different from the graph consisting of a single edge connecting Eve and Frank.)

7.1.7 Formulate the following assertion as a theorem about graphs, and prove it: At every party one can find two people who know the same number of other people (like Bob and Eve in our first example).

It will be instructive to give another proof of the theorem formulated in the last section. This will hinge on the answer to the following question: How many edges does a graph have? This can be answered easily if we think back to the problem of counting handshakes: For each node, we count the edges that leave that node (this is the degree of the node). If we sum these numbers, we count every edge twice. So dividing the sum by two, we get the number of edges. Let us formulate this observation as a theorem:

Theorem 7.1.2 *The sum of degrees of all nodes in a graph is twice the number of edges.*

In particular, we see that the sum of degrees in any graph is an even number. If we omit the even terms from this sum, we still get an even number. So the sum of odd degrees is even. But this is possible only if the number of odd degrees is even (since the sum of an odd number of odd numbers is odd). Thus we have obtained a new proof of Theorem 7.1.1.

7.2   Paths, Cycles, and Connectivity

Let us get acquainted with some special kinds of graphs. The simplest graphs are the *edgeless graphs*, having any number of nodes but no edges.

We get another very simple kind of graphs if we take n nodes and connect any two of them by an edge. Such a graph is called a *complete graph* (or a *clique*). A complete graph with n nodes is denoted by K_n. It has $\binom{n}{2}$ edges (recall Exercise 7.1.3).

If we think of a graph as representing some kind of relation, then it is clear that we could just as well represent the relation by connecting two nodes if they are not related. So for every graph G, we can construct another graph $\overline{G}$ that has the same node set but in which two nodes are connected precisely if they are *not* connected in the original graph G. The graph $\overline{G}$ is called the *complement* of G.

If we take n nodes and connect one of them to all the others, we get a *star*. This star has $n - 1$ edges.

Let us draw n nodes in a row and connect the consecutive ones by an edge. This way we obtain a graph with $n - 1$ edges, which is called a *path*.

The first and last nodes in the row are called the *endpoints* of the path. If we also connect the last node to the first, we obtain a *cycle* (or *circuit*). The number of edges in a path or cycle is called its *length*. A cycle of length k is often called a k-cycle. Of course, we can draw the same graph in many other ways, placing the nodes elsewhere, and we may get edges that intersect (Figure 7.5).

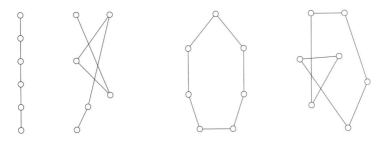

FIGURE 7.5. Two paths and two cycles

A graph H is called a *subgraph* of a graph G if it can be obtained from G by deleting some of its edges and nodes (of course, if we delete a node we automatically delete all the edges that connect it to other nodes).

7.2.1 Find all complete graphs, paths, and cycles among the graphs in Figures 7.1–7.5.

7.2.2 How many subgraphs does an edgeless graph on n nodes have? How many subgraphs does a triangle have?

7.2.3 Find all graphs that are paths or cycles and whose complements are also paths or cycles.

A key notion in graph theory is that of a *connected* graph. It is intuitively clear what this should mean, but it is also easy to formulate the property as follows: A graph G is connected if every two nodes of the graph are connected by a path in G. To be more precise: A graph G is connected if for every two nodes u and v, there exists a path with endpoints u and v that is a subgraph of G (Figure 7.6).

It will be useful to include a little discussion of this notion. Suppose that nodes a and b are connected by a path P in our graph. Also suppose that nodes b and c are connected by a path Q. Can a and c be connected by a path? The answer seems to be obviously "yes," since we can just go from a to b and then from b to c. But there is a difficulty: Concatenating (joining together) the two paths may not yield a path from a to c, since P and Q may intersect each other (Figure 7.7). But we can construct a path from a to c easily: Let us follow the path P to its first common node d with Q; then let us follow Q to c. Then the nodes we traversed are all distinct. Indeed,

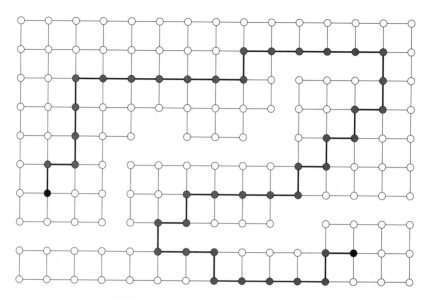

FIGURE 7.6. A path in a graph connecting two nodes

the nodes on the first part of our walk are distinct because they are nodes of the path P; similarly, the nodes on the second part are distinct because they are nodes of the path Q; finally, any node of the first part must be distinct from any node of the second part (except, of course, the node d), because d is the *first* common node of the two paths and so the nodes of P that we passed through before d are not nodes of Q at all. Hence the nodes and edges we have traversed form a path from a to c as claimed.[2]

A *walk* in a graph G is a sequence of nodes $v_0, v_1, \ldots, v_k$ such that v_0 is adjacent to v_1, which is adjacent to v_2, which is adjacent to v_3, etc.; any two consecutive nodes in the sequence must be connected by an edge. This sounds almost like a path: The difference is that a walk may pass through the same node several times, while a path must go through different nodes. Informally, a walk is a "path with repetition"; more correctly, a path is a walk without repetition. Even the first and last nodes of the walk may be the same; in this case, we call it a *closed walk*. The shortest possible walk consists of a single node v_0 (this is closed). If the first node v_0 is different from the last node v_k, then we say that this walk *connects* nodes v_0 and v_k.

---

[2]We have given more details of this proof than was perhaps necessary. One should note, however, that when arguing about paths and cycles in graphs, it is easy to draw pictures (on the paper or mentally) that make implicit assumptions and are therefore misleading. For example, when joining together two paths, one's first mental image is a single (longer) path, which may not be the case.

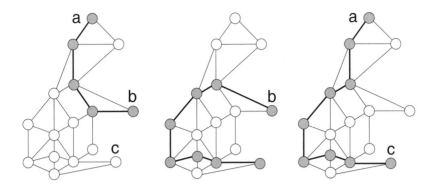

FIGURE 7.7. Selecting a path from a to c, given a path from a to b and a path from b to c.

Is there a difference between connecting two nodes by a walk and connecting them by a path? Not really: If two nodes can be connected by a walk, then they can also be connected by a path. Sometimes it is more convenient to use paths, sometimes, to use walks (see Exercise 7.2.6).

Let G be a graph that is not necessarily connected. G will have connected subgraphs; for example, the subgraph consisting of a single node (and no edge) is connected. A *connected component* H is a maximal subgraph that is connected; in other words, H is a connected component if it is connected but every other subgraph of G that contains H is disconnected. It is clear that every node of G belongs to some connected component. It follows by Exercise 7.2.7 that different connected components of G have no node in common (otherwise, their union would be a connected subgraph containing both of them). In other words, every node of G is contained in a unique connected component.

7.2.4 Is the proof as given above valid if (a) the node a lies on the path Q; (b) the paths P and Q have no node in common except b?

7.2.5   (a) We delete an edge e from a connected graph G. Show by an example that the remaining graph may not be connected.

  (b) Prove that if we assume that the deleted edge e belongs to a cycle that is a subgraph of G, then the remaining graph is connected.

7.2.6 Let G be a graph and let u and v be two nodes of G.

  (a) Prove that if there is a walk in G from u to v, then G contains a path connecting u and v.

  (b) Use part (a) to give another proof of the fact that if G contains a path connecting a and b, and also a path connecting b and c, then it contains a path connecting a and c.

7.2.7 Let G be a graph, and let $H_1 = (V_1, E_1)$ and $H_2 = (V_2, E_2)$ be two subgraphs of G that are connected. Assume that H_1 and H_2 have at least one node in common. Form their union, i.e., the subgraph $H = (V', E')$, where $V' = V_1 \cup V_2$ and $E' = E_1 \cup E_2$. Prove that H is connected.

7.2.8 Determine the connected components of the graphs constructed in Exercise 7.1.4.

7.2.9 Prove that no edge of G can connect nodes in different connected components.

7.2.10 Prove that a node v is a node of the connected component of G containing node u if and only if g contains a path connecting u to v.

7.2.11 Prove that a graph with n nodes and more than $\binom{n-1}{2}$ edges is always connected.

7.3   Eulerian Walks and Hamiltonian Cycles

Perhaps the oldest result in graph theory was discovered by Leonhard Euler, the greatest mathematician of the eighteenth century.

FIGURE 7.8. Leonhard Euler 1707–1783

It started with a recreational challenge that the citizens of Königsberg (today, Kaliningrad) raised. The city was divided into four districts by branches of the river Pregel (Figure 7.9), which were connected by seven bridges. It was nice to walk around, crossing these bridges, and so the question arose, is it possible to take a walk so that one crosses every bridge exactly once?

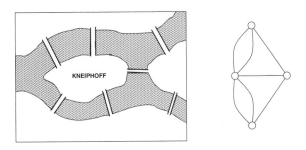

FIGURE 7.9. The bridges of Königsberg in Euler's time, and the graph modeling them.

Euler published a paper in 1736 in which he proved that such a walk was impossible. The argument is quite simple. Suppose that there is such a walk. Consider any of the four parts of the town, say the island Kneiphoff, and suppose that our walk does not start here. Then at some point in time, we enter the island by crossing a bridge; somewhat later, we leave it through another bridge (by the rules of the walk). Then we enter it again through a third bridge, then leave it through a fourth, then enter it through the fifth, then.... We cannot leave the island (at least not as part of the walk), since we have used up all the bridges that lead to it. We must end our walk on the island.

So we must either start or end our walk on the island. This is OK—the rules don't forbid it. The trouble is that we can draw the same conclusion for any of the other three districts of the town. The only difference is that instead of five bridges, these districts are connected to the rest of the town by only three bridges; so if we don't start there, we get stuck there at the second visit, not the third.

But now we are in trouble: we cannot start or end the walk in each of the four districts! This proves that no walk can cross every bridge exactly once.

Euler remarked that one could reach this conclusion by making an exhaustive list of all possible routes, and checking that none of them can be completed as required; but this would be impractical due to the large number of possibilities. More significantly, he formulated a general criterion by which one could decide for every city (no matter how many islands and bridges it had) whether one could take a walk crossing every bridge exactly once.

Euler's result is generally regarded as the first theorem of graph theory. Of course, Euler did not have the terminology of graphs (which was not to appear for more than a century), but we can use it to state Euler's theorem.

Let G be a graph; for the following discussion, we allow parallel edges, i.e., several edges connecting the same pair of nodes. A *walk* in such a graph is a bit more difficult to define. It consists of a sequence of nodes again such that any two consecutive nodes are connected by an edge; but if there are several edges connecting these consecutive nodes, we also have to specify which of these edges is used to move from one node to the next. So formally, a walk in a graph with parallel edges is a sequence $v_0, e_1, v_1, e_2, v_2, \ldots, v_{k-1}, e_k, v_k$, where $v_0, v_1, \ldots, v_k$ are nodes, $e_1, e_2, \ldots, e_k$ are edges, and edge e_i connects nodes v_{i-1} and v_i $(i = 1, 2, \ldots, k)$.

An *Eulerian walk* is a walk that goes through every edge exactly once (the walk may or may not be closed; see Figure 7.10). To see how to cast the problem of the Königsberg bridges into this language, let us represent each district by a node and draw an edge connecting two nodes for every bridge connecting the two corresponding districts. We get the little graph on the right hand side of Figure 7.9. A walk in the town corresponds to a walk in this graph (at least, if only crossing the bridges matters), and a walk that crosses every bridge exactly once corresponds to an Eulerian walk.

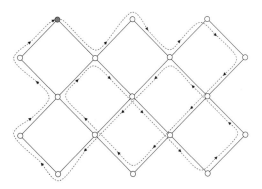

FIGURE 7.10. An Eulerian walk in a graph.

Recast in this language, Euler's criteria are stated in the following theorem.

Theorem 7.3.1 (a) *If a connected graph has more than two nodes with odd degree, then it has no Eulerian walk.*

(b) *If a connected graph has exactly two nodes with odd degree, then it has an Eulerian walk. Every Eulerian walk must start at one of these and end at the other one.*

(c) *If a connected graph has no nodes with odd degree, then it has an Eulerian walk. Every Eulerian walk is closed.*

Proof. Euler's argument above gives the following: *If a node v has odd degree, then every Eulerian walk must either start or end at v.* Similarly, we can see that *if a node v has even degree, then every Eulerian walk either starts and ends at v, or starts and ends somewhere else.* This observation immediately implies (a), as well as the second assertions in (b) and (c).

To finish the proof, we have to show that if a connected graph has 0 or 2 nodes with odd degree, then it has an Eulerian walk. We describe the proof in the case where there is no node of odd degree (part (c)); the other case is left to the reader as Exercise 7.3.14.

Let v be any node. Consider a closed walk starting and ending at v that uses every edge at most once. Such a walk exists. For example, we can take the walk consisting of the node v only. But we don't want this very short walk; instead, we consider a longest closed walk W starting at v, using every edge at most once.

We want to show that this walk W is Eulerian. Suppose not. Then there is at least one edge e that is not used by W. We claim that we can choose this edge so that W passes through at least one of its endpoints. Indeed, if

p and q are the endpoints of e and W does not pass through them, then we take a path from p to v (such a path exists since the graph is connected), and look at the first node r on this path that is also on the walk W (Figure 7.11(a)). Let $e' = sr$ be the edge of the path just before r. Then W does

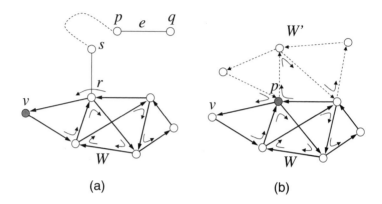

FIGURE 7.11. (a) Finding an edge not in W but meeting W. (b) Combining W and W'.

not pass through e (because it does not pass through s), so we can replace e by e', which has one endpoint on W.

So let e be an edge that is not used by W but has an endpoint p that is used by W. Then we start a new walk W' at p. We start through e, and continue walking as we please, only taking care that (i) we don't use the edges of W, and (ii) we don't use any edge twice.

Sooner of later we get stuck, but where? Let u be the node where we get stuck, and suppose that $u \neq p$. Node u has even degree; W uses up an even number of edges incident with u; every previous visit of the new walk to this node used up two edges (in and out); our last entrance used up one edge; so we have an odd number of edges that are edges neither of W nor of W'. But this means that we can continue our walk!

So the only node we can get stuck in is node p. This means that W' is a closed walk. Now we take a walk as follows. Starting at v, we follow W to p; then follow W' all the way through, so that eventually we get back to p; then follow W to its end at v (Figure 7.11(b)). This new walk starts and ends at v, uses every edge at most once, and is longer than W, which is a contradiction.   □

Euler's result above is often formulated as follows: *A connected graph has a closed Eulerian walk if and only if every node has even degree.*

7.3.1 Which of the graphs in Figure 7.12 have an Eulerian walk? Which of them have a closed Eulerian walk? Find an Eulerian walk if it exists.

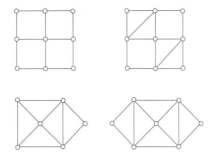

FIGURE 7.12. Which of these graphs has an Eulerian walk?

7.3.2 When does a connected graph contain two walks such that every edge is used by exactly one of them, exactly once?

A question similar to the problem of the Bridges of Königsberg was raised by another famous mathematician, the Irish William R. Hamilton, in 1856. A *Hamiltonian cycle* is a cycle that contains all nodes of a graph. The Hamilton cycle problem is the problem of deciding whether or not a given graph has a Hamiltonian cycle.

Hamiltonian cycles sound quite similar to Eulerian walks: Instead of requiring that every edge be used exactly once, we require that every node be used exactly once. But much less is known about them than about Eulerian walks. Euler told us how to decide whether a given graph has an Eulerian walk; but no efficient way is known to check whether a given graph has a Hamiltonian cycle, and no useful necessary and sufficient condition for the existence of a Hamiltonian cycle is known. If you solve Exercise 7.3.3, you'll get a feeling about the difficulty of the Hamiltonian cycle problem.

7.3.3 Decide whether the graphs in Figure 7.13 have a Hamiltonian cycle.

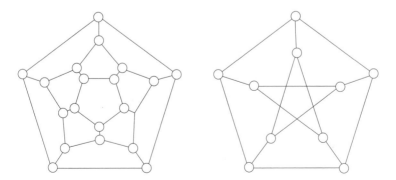

FIGURE 7.13. Two famous graphs: the dodecahedron graph (cf. Chapter 12) and the Petersen graph.

Review Exercises

7.3.4 Draw all graphs on 5 nodes in which every node has degree at most 2.

7.3.5 Does there exists a graph with the following degrees: (a) $0, 2, 2, 2, 4, 4, 6$; (b) $2, 2, 3, 3, 4, 4, 5$.

7.3.6 Draw the graphs representing the bonds between atoms in (a) a water molecule; (b) a methane molecule; (c) two water molecules.

7.3.7 At a party there were 7 boys and 6 girls. Every boy danced with every girl. Draw the graph representing the dancing. How many edges does it have? What are its degrees?

7.3.8 How many subgraphs does a 4-cycle have?

7.3.9 Prove that at least one of G and $\overline{G}$ is connected.

7.3.10 Let G be a connected graph with at least two nodes. Prove that it has a node such that if this node is removed (along with all edges incident with it), the remaining graph is connected.

7.3.11 Let G be a connected graph that is not a path. Prove that it has at least three vertices such that if any of them is removed, the remaining graph is still connected.

7.3.12 Let G be a connected graph in which every pair of edges have an endpoint in common. Show that G is either a star or a K_3.

7.3.13 There are $(m-1)n + 1$ people in a room. Show that either there are m people who mutually do not know each other, or there is a person who knows at least n others.

7.3.14 Prove part (b) of Theorem 7.3.1.

7.3.15 Theorem 7.3.1 talks about connected graphs. Which disconnected graphs have an Eulerian walk?

8
Trees

8.1 How to Define Trees

We have met trees when we were studying enumeration problems; now we take a look at them as graphs. A graph $G = (V, E)$ is called a *tree* if it is connected and contains no cycle as a subgraph. The simplest tree has one node and no edges. The second simplest tree consists of two nodes connected by an edge. Figure 8.1 shows a variety of other trees.

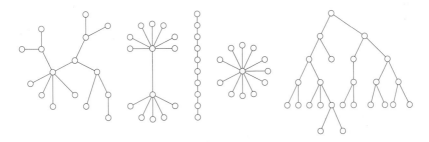

FIGURE 8.1. Five trees.

Note that the two properties defining trees work in opposite directions: Connectedness means that the graph cannot have "too few" edges, while the exclusion of cycles means that it cannot have "too many." To be more precise, if a graph is connected, then if add a new edge to it, it remains connected (while if we delete an edge, it may become disconnected). If a graph contains no cycle, then if we delete any edge, the remaining graph will still not contain a cycle (while adding a new edge may create a cycle). The following theorem shows that trees can be characterized as "minimally connected" graphs as well as "maximally cycle-free" graphs.

Theorem 8.1.1 *(a) A graph G is a tree if and only if it is connected, but deleting any of its edges results in a disconnected graph.*

(b) A graph G is a tree if and only if it contains no cycles, but adding any new edge creates a cycle.

Proof. We prove part (a) of this theorem; the proof of part (b) is left as an exercise.

First, we have to prove that if G is a tree then it satisfies the condition given in the theorem. It is clear that G is connected (by the definition of a tree). We want to prove that if we delete any edge, it cannot remain connected. The proof is indirect: Assume that when the edge uv is deleted from a tree G, the remaining graph G' is connected. Then G' contains a path P connecting u and v. But then, if we put the edge uv back, the path P and the edge uv will form a cycle in G, which contradicts the definition of trees.

Second, we have to prove that if G satisfies the condition given in the theorem, then it is a tree. It is clear that G is connected, so we only have to argue that G does not contain a cycle. Again by an indirect argument, assume that G does contain a cycle C. Then deleting any edge of C, we obtain a connected graph (Exercise 7.2.5). But this contradicts the condition in the theorem.   □

Consider a connected graph G on n nodes, and an edge e of G. If we delete e, the remaining graph may or may not remain connected. If it is disconnected, then we call e a *cut-edge*. Part (a) of Theorem 8.1.1 implies that every edge of a tree is a cut-edge.

If we find an edge that is not a cut-edge, delete it. Go on deleting edges until a graph is obtained that is still connected, but deleting any edge from it leaves a disconnected graph. By part (a) of Theorem 8.1.1, this is a tree, with the same node set as G. A subgraph of G with the same node set that is a tree is called a *spanning tree* of G. The edge deletion process above can, of course, be carried out in many ways, so a connected graph can have many different spanning trees.

Rooted trees. Often, we use trees that have a special node, which we call the *root*. For example, trees that occurred in counting subsets or permutations were built starting with a given node.

We can take any tree, select any of its nodes, and call it a root. A tree with a specified root is called a *rooted tree*.

Let G be a rooted tree with root r. Given any node v different from r, we know from Exercise 8.1.3 below that the tree contains a unique path connecting v to r. The node on this path next to v is called the *father* of

v. The other neighbors of v are called the *sons* of v. The root r does not have a father, but all its neighbors are called its sons.

We now make a basic genealogical assertion: *Every node is the father of its sons.* Indeed, let v be any node, and let u be one of its sons. Consider the unique path P connecting v to r. The node u cannot lie on P: It cannot be the first node after v, since then it would be the father of v, and not its son; and it cannot be a later node, since then in going from v to u on the path P and then back to v on the edge uv we would traverse a cycle. But this implies that in adding the node u and the edge uv to P we get a path connecting u to r. Since v is the first node on this path after u, it follows that v is the father of u. (Is this argument valid when $v = r$? Check!)

We have seen that every node different from the root has exactly one father. A node can have any number of sons, including zero. A node with no sons is called a *leaf*. In other words, a leaf is a node with degree 1, different from r.

8.1.1 Prove part (b) of Theorem 8.1.1.

8.1.2 Prove that connecting two nodes u and v in a graph G by a new edge creates a new cycle if and only if u and v are in the same connected component of G.

8.1.3 Prove that in a tree, every two nodes can be connected by a *unique* path. Conversely, prove that if a graph G has the property that every two nodes can be connected by a path, and there is only one connecting path for each pair, then the graph is a tree.

8.2   How to Grow Trees

The following is one of the most important properties of trees.

Theorem 8.2.1 *Every tree with at least two nodes has at least two nodes of degree 1.*

Proof. Let G be a tree with at least two nodes. We prove that G has a node of degree 1, and leave it to the reader as an exercise to prove that it has at least one more. (A path has only two such nodes, so this is the best possible we can claim.)

Let us start from any node v_0 of the tree and take a walk (climb?) on the tree. Let's say we never want to turn back from a node on the edge through which we entered it; this is possible unless we get to a node of degree 1, in which case we stop and the proof is finished.

So let's argue that this must happen sooner or later. If not, then eventually we must return to a node we have already visited; but then the nodes and edges we have traversed between the two visits form a cycle. This contradicts our assumption that G is a tree and hence contains no cycle. □

8.2.1 Apply the argument above to find a second node of degree 1.

A real tree grows by developing new twigs again and again. We show that graph-trees can be grown in the same way. To be more precise, consider the following procedure, which we call the *Tree-growing Procedure*:

— *Start with a single node.*

— *Repeat the following any number of times: If you have any graph G, create a new node and connect it by a new edge to any node of G.*

Theorem 8.2.2 *Every graph obtained by the Tree-growing Procedure is a tree, and every tree can be obtained this way.*

Proof. The proof of this is again rather straightforward, but let us go through it, if only to gain practice in arguing about graphs.

First, consider any graph that can be obtained by this procedure. The starting graph is certainly a tree, so it suffices to argue that we never create a nontree; in other words, if G is a tree, and G' is obtained from G by creating a new node v and connecting it to a node u of G, then G' is a tree. This is straightforward: G' is connected, since any two "old" nodes can be connected by a path in G, while v can be connected to any other node w by first going to u and then connecting u to w. Moreover, G cannot contain a cycle: v has degree 1, and so no cycle can go through v, but a cycle that does not go through v would be a cycle in the old graph, which is supposed to be a tree.

Second, let's argue that every tree can be constructed this way. We prove this by induction on the number of nodes.[1] If the number of nodes is 1, then the tree arises by the construction, since this is the way we start. Assume that G is a tree with at least 2 nodes. Then by Theorem 8.2.1, G has a node of degree 1 (at least two nodes, in fact). Let v be a node with degree 1. Delete v from G, together with the edge with endpoint v, to get a graph G'.

We claim that G' is a tree. Indeed, G' is connected: Any two nodes of G' can be connected by a path in G, and this path cannot go through v, since v has degree 1. So this path is also a path in G'. Furthermore, G' does not contain a cycle since G does not.

---

[1] The first part of the proof is also an induction argument, even though it was not phrased as such.

By the induction hypothesis, every tree with fewer nodes than G arises by the construction; in particular, G' does. But then G arises from G' by one more iteration of the second step. This completes the proof of Theorem 8.2.2. □

Figure 8.2 shows how trees with up to 4 nodes arise by this construction. Note that there is a "tree of trees" here. The fact that the logical structure of this construction is a tree does not have anything to do with the fact that we are constructing trees: any iterative construction with free choices at each step results in a similar "descent tree".

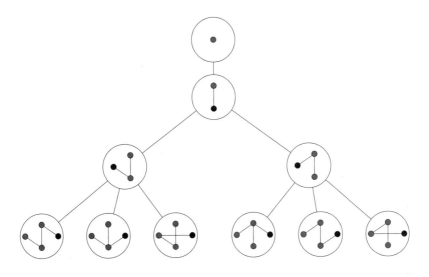

FIGURE 8.2. The descent tree of trees

The Tree-growing Procedure can be used to establish a number of properties of trees. Perhaps most important of these concerns the number of edges. How many edges does a tree have? Of course, this depends on the number of nodes; but surprisingly, it depends *only* on the number of nodes:

Theorem 8.2.3 *Every tree on n nodes has $n - 1$ edges.*

Proof. Indeed, we start with one more node (1) than edge (0), and at each step, one new node and one new edge are added, so this difference of 1 is maintained. □

8.2.2 Let G be a tree, which we consider as the network of roads in a medieval country, with castles as nodes. The king lives at node r. On a certain day, the lord of each castle sets out to visit the king. Argue carefully that soon after they have left their castles, there will be exactly one lord on each edge. Give a proof of Theorem 8.2.3 based on this.

8.2.3 If we delete a node v from a tree (together with all edges that end there), we get a graph whose connected components are trees. We call these connected components the *branches* at node v. Prove that every tree has a node such that every branch at this node contains at most half the nodes of the tree.

8.3 How to Count Trees?

We have counted all sorts of things in the first part of this book; now that we are familiar with trees, it is natural to ask: *How many trees are there on n nodes?*

Before attempting to answer this question, we have to clarify an important issue: when do we consider two trees different? There is more than one reasonable answer to this question. Consider the trees in Figure 8.3. Are they the same? One could say that they are; but then, if the nodes are, say, towns, and the edges represent roads to be built between them, then clearly the inhabitants of the towns will consider the two plans very different.

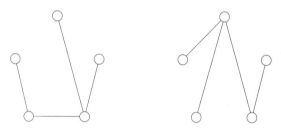

FIGURE 8.3. Are these trees the same?

So we have to define carefully when we consider two trees the same. The following are two possibilities:

— We fix the set of nodes, and consider two trees the same if the same pairs of nodes are connected in each. (This is the position the townspeople would take when they consider road construction plans.) In this case, it is advisable to give names to the nodes, so that we can distinguish them. It is convenient to use the numbers $0, 1, 2, \ldots, n-1$ as names (if the tree has n nodes). We express this by saying that the vertices of the tree are labeled by $0, 1, 2, \ldots n - 1$. Figure 8.4 shows a labeled tree. Interchanging the labels 2 and 4 (say) would yield a different labeled tree.

— We don't give names to the nodes, and consider two trees the same if we can rearrange the nodes of one so that we get the other tree. More exactly, we consider two trees the same (the mathematical term for this is *isomorphic*) if there exists a one-to-one correspondence between the nodes of the first tree and the nodes of the second tree such that two nodes in the first tree that are connected by an edge correspond to nodes in the second tree that are connected by an edge, and vice versa. If we speak about *unlabeled trees*, we mean that we don't distinguish isomorphic trees from each other. For example, all paths on n nodes are the same as unlabeled trees.

So we can ask two questions: How many labeled trees are there on n nodes? and how many unlabeled trees are there on n nodes? These are really two different questions, and we have to consider them separately.

8.3.1 Find all unlabeled trees on $2, 3, 4$, and 5 nodes. How many labeled trees do you get from each? Use this to find the number of labeled trees on $2, 3, 4$, and 5 nodes.

8.3.2 How many labeled trees on n nodes are stars? How many are paths?

The number of labeled trees. For the case of labeled trees, there is a very nice solution.

Theorem 8.3.1 (Cayley's Theorem) *The number of labeled trees on n nodes is n^{n-2}.*

The formula is elegant, but the surprising fact about it is that it is quite difficult to prove! It is substantially deeper than any of the previous formulas for the number of this and that. There are various ways to prove it, but each uses some deeper tool from mathematics or a deeper idea. We'll give a proof that is perhaps best understood by first discussing a quite different question in computer science: how to store trees.

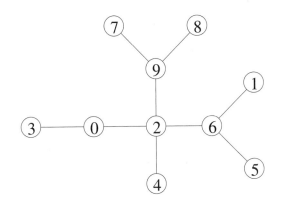

FIGURE 8.4. A labeled tree

8.4   How to Store Trees

Suppose that you want to store a labeled tree, say the tree in Figure 8.4, in a computer. How would you do this? Of course, the answer depends on what you need to store the tree for, what information about it you want to retrieve and how often, etc. Right now, we are concerned only with the amount of memory we need. We want to store the tree so that it occupies the least amount of memory.

Let's try some simple solutions.

(a) Suppose that we have a tree G with n nodes. One thing that comes to mind is to make a big table, with n rows and n columns, and put (say) the number 1 in the jth position of the ith row if nodes i and j are connected by an edge, and the number 0, if they are not. It will be convenient to place the node labeled 0 last, so it corresponds to the 10th row and to the 10th column:

$$
\begin{array}{cccccccccc}
0 & 0 & 0 & 0 & 0 & 1 & 0 & 0 & 0 & 0 \\
0 & 0 & 0 & 1 & 0 & 1 & 0 & 0 & 0 & 1 \\
0 & 0 & 0 & 0 & 0 & 0 & 0 & 0 & 0 & 1 \\
0 & 1 & 0 & 0 & 0 & 0 & 0 & 0 & 0 & 0 \\
0 & 0 & 0 & 0 & 0 & 1 & 0 & 0 & 0 & 0 \\
1 & 1 & 0 & 0 & 1 & 0 & 0 & 0 & 0 & 0 \\
0 & 0 & 0 & 0 & 0 & 0 & 0 & 0 & 1 & 0 \\
0 & 0 & 0 & 0 & 0 & 0 & 0 & 0 & 1 & 0 \\
0 & 0 & 0 & 0 & 0 & 0 & 1 & 1 & 0 & 0 \\
0 & 1 & 1 & 0 & 0 & 0 & 0 & 0 & 0 & 0 \\
\end{array}
\tag{8.1}
$$

This method of storing the tree can, of course, be used for any graph (it is called the *adjacency matrix* of the graph, just to mention its name). It is often very useful, but at least for trees, it is very wasteful. We need one bit to store each entry of this table, so this takes n^2 bits. We can save a little by noticing that it is enough to store the part below the diagonal, since the diagonal is always 0 and the other half of the table is just the reflection of the half below the diagonal. But this is still $(n^2 - n)/2$ bits.

(b) We fare better if we specify each tree by listing all its edges. We can specify each edge by its two endpoints. It will be convenient to arrange this list in an array whose columns correspond to the edges. For example, the tree in Figure 8.4 can be encoded by

$$
\begin{array}{ccccccccc}
7 & 8 & 9 & 6 & 3 & 0 & 2 & 6 & 6 \\
9 & 9 & 2 & 2 & 0 & 2 & 4 & 1 & 5 \\
\end{array}
$$

Instead of a table with n rows, we get a table just with two rows. We pay a little for this: Instead of just 0 and 1, the table will contain integers between 0 and $n - 1$. But this is certainly worth it: Even if we count bits, to write down the label of a node takes $\log_2 n$ bits, so the whole table occupies only $2n \log_2 n$ bits, which is much less than $(n^2 - n)/2$ if n is large.

There is still a lot of free choice here, which means that the same tree may be encoded in different ways: We have freedom in choosing the order of the edges, and also in choosing the order in which the two endpoints of an edge are listed. We could agree on some arbitrary conventions to make the code well defined (say, listing the two endnodes of an edge in increasing order, and then the edges in increasing order of their first endpoints, breaking ties according to the second endpoints); but it will be more useful to do this in a way that also allows us to save more memory.

(c) **The father code.** From now on, the node with label 0 will play a special role; we'll consider it the "root" of the tree. Then we can list the two endnodes of an edge by listing the endpoint further from the root first, and then the endpoint nearer to the root second. So for every edge, the node written below is the father of the node written above. For the order in which we list the edges, let us take the order of their first nodes. For the tree in Figure 8.4, we get the table

$$
\begin{array}{ccccccccc}
1 & 2 & 3 & 4 & 5 & 6 & 7 & 8 & 9 \\
6 & 0 & 0 & 2 & 6 & 2 & 9 & 9 & 2
\end{array}
$$

Do you notice anything special about this table? The first row consists of the numbers $1, 2, 3, 4, 5, 6, 7, 8, 9$, in this order. Is this a coincidence? Well, the order is certainly not (we ordered the edges by the increasing order of their first endpoints). The root 0 does not occur, since it is not the son of any other node. But why do we get every other number exactly once? After a little reflection, this should also be clear: If a node occurs in the first row, then its father occurs below it. Since a node has only one father, it can occur only once. Since every node other than the root has a father, every node other than the root occurs in the first row.

Thus we know in advance that if we have a tree on n nodes, and write up the array using this method, then the first row will consist of $1, 2, 3, \ldots, n-1$. So we may as well suppress the first row without losing any information; it suffices to store the second row. So we can specify the tree by a sequence of $n-1$ numbers, each between 0 and $n-1$. This takes $(n-1)\lceil \log_2 n \rceil$ bits.

This coding is not optimal, in the sense that not every "code" gives a tree (see Exercise 8.4.1). But we'll see that this method is already nearly optimal.

8.4.1 Consider the following "codes": $(0, 1, 2, 3, 4, 5, 6, 7)$; $(7, 6, 5, 4, 3, 2, 1, 0)$; $(0, 0, 0, 0, 0, 0, 0, 0)$; $(2, 3, 1, 2, 3, 1, 2, 3)$. Which of these are "father codes" of trees?

8.4.2 Prove, based on the "father code" method of storing trees, that the number of labeled trees on n nodes is at most n^{n-1}.

(d) Now we describe a procedure, called the *Prüfer code*, that will assign to any n-point labeled tree a sequence of length $n-2$, not $n-1$, consisting of the numbers $0, \ldots, n-1$. The gain is small, but important: we'll show that every such sequence corresponds to a tree. Thus we will establish a *bijection*, a one-to-one correspondence, between labeled trees on n nodes and sequences of length $n-2$, consisting of numbers $0, 1, \ldots, n-1$. Since the number of such sequences is n^{n-2}, this will also prove Cayley's Theorem.

The Prüfer code can be considered as a refinement of method (c). We still consider 0 as the root, we still order the two endpoints of an edge so that the son comes first, but we order the edges (the columns of the array) not by the magnitude of their first endpoint but a little differently, more closely related to the tree itself.

So again, we construct a table with two rows, whose columns correspond to the edges, and each edge is listed so that the node farther from 0 is on the top, its father on the bottom. The issue is the order in which we list the edges.

Here is the rule for this order: We look for a node of degree 1, different from 0, with the smallest label, and write down the edge with this endnode. In our example, this means that we write down $\frac{1}{6}$. Then we delete this node and edge from the tree, and repeat: We look for the endnode with smallest label, different from 0, and write down the edge incident with it. In our case, this means adding a column $\frac{3}{0}$ to the table. Then we delete this node and edge, etc. We go until all edges are listed. The array we get is called the *extended Prüfer code* of the tree (we call it extended because, as we'll see, we need only a part of it as the "real" Prüfer code). The extended Prüfer code of the tree in Figure 8.4 is:

$$
\begin{array}{ccccccccc}
1 & 3 & 4 & 5 & 6 & 7 & 8 & 9 & 2 \\
6 & 0 & 2 & 6 & 2 & 9 & 9 & 2 & 0
\end{array}
$$

Why is this any better than the "father code"? One little observation is that the last entry in the second row is now always 0, since it comes from the last edge and since we never touched the node 0, this last edge must be incident with it. But we have paid a lot for this, it seems: It is no longer clear that the first row is superfluous; it still consists of the numbers $1, 2, \ldots, n-1$, but now they are not in increasing order.

The key lemma is that the first row is determined by the second:

Lemma 8.4.1 *The second row of an extended Prüfer code determines the first.*

Let us illustrate the proof of the lemma by an example. Suppose that somebody gives us the second row of an extended Prüfer code of a labeled tree on 8 nodes; say, $2\,4\,0\,3\,3\,1\,0$ (we have one edges fewer than nodes, so the second row consists of 7 numbers, and as we have seen, it must end with a 0). Let us figure out what the first row must have been.

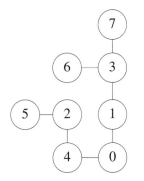

FIGURE 8.5. A tree reconstructed from its Prüfer code

How does the first row start? Remember that this is the node that we delete in the first step; by the rule of constructing the Prüfer code, this is the node with degree 1 with smallest label. Could this node be node 1? No, because then we would have to delete it in the first step, and it could no longer occur, but it does. By the same token, no number occurring in the second row could be a leaf of the tree at the beginning. This rules out $2, 3$, and 4.

What about 5? It does not occur in the second row; does this mean that it is a leaf of the original tree? The answer is yes; otherwise, 5 would have been the father of some other node, and it would have been written in the second row when this other node was deleted. Thus 5 was a leaf with smallest label, and the first row of the extended Prüfer code must start with 5.

Let's try to figure out the next entry in the first row, which is, as we know, the leaf with smallest label of the tree after 5 was deleted. The node 1 is still ruled out, since it occurs later in the second row ; but 2 does not occur again, which means (by the same argument as before) that 2 was the leaf with smallest label after 5 was deleted. Thus the second entry in the first row is 2.

Similarly, the third entry must be 4, since all the smaller numbers either occur later or have been used already. Continuing in a similar fashion, we get that the full array must have been

$$
\begin{array}{ccccccc}
5 & 2 & 4 & 6 & 7 & 3 & 1 \\
2 & 4 & 0 & 3 & 3 & 1 & 0
\end{array}
$$

This corresponds to the tree in Figure 8.5

Proof. [of Lemma 8.4.1] The considerations above are completely general, and can be summed up as follows:

> *Each entry in the first row of the extended Prüfer code is the smallest integer that does not occur in the first row before it, nor in the second row below or after it.*

Indeed, when this entry (say, the k-th entry in the first row) was recorded, the nodes before it in the first row were deleted (together with the edges corresponding to the first $k - 1$ columns). The remaining entries in the second row are exactly those nodes that are fathers at this time, which means that they are not leaves.

This describes how the first row can be reconstructed from the second. □

So we don't need the full extended Prüfer code to store the tree; it suffices to store the second row. In fact, we know that the last entry in the second row is 0, so we don't have to store this either. The sequence consisting of the first $n - 2$ entries of the second row is called the *Prüfer code* of the tree. Thus the Prüfer code is a sequence of length $n - 2$, each entry of which is a number between 0 and $n - 1$.

This is similar to the father code, just one shorter; not much gain here for all the work. But the beauty of the Prüfer code is that it is optimal, in the sense that

> *every sequence of numbers between 0 and $n - 1$, of length $n - 2$, is a Prüfer code of some tree on n nodes.*

This can be proved in two steps. First, we extend this sequence to a table with two rows: We add a 0 at the end, and then write above each entry in the first row the smallest integer that does not occur in the first row before it, nor in the second row below or after it (note that it is always possible to find such an integer: the condition excludes at most $n - 1$ values out of n).

Now this table with two rows is the Prüfer code of a tree. The proof of this fact, which is no longer difficult, is left to the reader as an exercise.

8.4.3 Complete the proof.

Let us sum up what the Prüfer code gives. First, it proves Cayley's theorem. Second, it provides a theoretically most efficient way of encoding trees. Each Prüfer code can be considered as a natural number written in the base-n number system; in this way, we associate a "serial number" between 0 and $n^{n-2} - 1$ with each n-point labeled tree. Expressing these serial numbers in base two, we get a code using 0–1 sequences of length at most $\lceil (n - 2) \log_2 n \rceil$.

As a third use of the Prüfer code, let's suppose that we want to write a program that generates a random labeled tree on n nodes in such a way that all trees occur with the same probability. This is not easy from scratch; but the Prüfer code gives an efficient solution. We just have to generate $n - 2$ independent random integers between 0 and $n - 1$ (most programming languages have a statement for this) and then "decode" this sequence as a tree, as described.

8.5   The Number of Unlabeled Trees

The number of unlabeled trees on n nodes, usually denoted by T_n, is even more difficult to handle. No simple formula like Cayley's theorem is known for this number. Our goal is to get a rough idea of how large this number is.

There is only one unlabeled tree on $1, 2$, or 3 nodes; there are two on 4 nodes (the path and the star). There are 3 on 5 nodes (the star, the path, and the tree in Figure 8.3. These numbers are much smaller than the number of labeled trees with these numbers of nodes, which are $1, 1, 3, 16$, and 125 by Cayley's theorem.

It is of course clear that the number of unlabeled trees is less than the number of labeled trees; every unlabeled tree can be labeled in many ways. How many ways? If we draw an unlabeled tree, we can label its nodes in $n!$ ways. The labeled trees we get this way are not necessarily all different. For example, if the tree is a star, then no matter how we permute the labels of the leaves, we get the same labeled tree. So an unlabeled star yields n labeled stars.

But at least we know that each labeled tree can be labeled in at most $n!$ ways. Since the number of labeled trees is n^{n-2}, it follows that the number of unlabeled trees is at least $n^{n-2}/n!$. Using Stirling's formula (Theorem 2.2.1), we see that this number is about $e^n/n^{5/2}\sqrt{2\pi}$.

This number is much smaller than the number of labeled trees, n^{n-2}, but of course it is only a *lower bound* on the number of unlabeled trees. How can we obtain an *upper bound* on this number? If we think in terms of storage, the issue is, can we store an unlabeled tree more economically than labeling its nodes and then storing it as a labeled tree? Very informally, how should we describe a tree if we want only the "shape" of it, and don't care which node gets which label?

Take an n-point tree G, and specify one of its leaves as its "root." Next, draw G in the plane without crossing edges; this can always be done, and we almost always draw trees this way.

Now we imagine that the edges of the tree are walls, perpendicular to the plane. Starting at the root, walk around this system of walls, keeping the wall always to your right. We'll call walking along an edge a *step*. Since there are $n - 1$ edges, and each edge has two sides, we'll make $2(n - 1)$ steps before returning to the root (Figure 8.6).

Each time we make a step *away* from the root (i.e., a step from a father to one of its sons), we write down a 1; each time we make a step *toward* the root, we write down a 0. This way we end up with a sequence of length $2(n - 1)$, consisting of 0's and 1's. We call this sequence the *planar code* of the (unlabeled) tree. The planar code of the tree in Figure 8.6 is 1111100100011011010000.

Now this name already indicates that the planar code has the following important property:

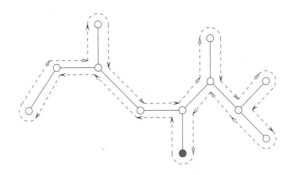

FIGURE 8.6. Walking around a tree.

Every unlabeled tree is uniquely determined by its planar code.

Let us illuminate the proof of this by assuming that the tree is covered by snow, and we have only its code. We ask a friend of ours to walk around the tree just as above, and uncover the walls, and we look at the code in the meanwhile. What do we see? Any time we see a 1, he walks along a wall away from the root, and he cleans the snow from it. We see this as growing a new twig. Any time we see a 0, he walks back, along an edge already uncovered, toward the root.

Now, this describes a perfectly good way to draw the tree: We look at the bits of the code one by one, while keeping the pen on the paper. Any time we see a 1, we draw a new edge to a new node (and move the pen to the new node). Any time we see a 0, we move the pen back by one edge toward the root. Thus the tree is indeed determined by its planar code.

Since the number of possible planar codes is at most $2^{2(n-1)} = 4^{n-1}$, we get that the number of unlabeled trees is at most this large. Summing up:

Theorem 8.5.1 *The number T_n of unlabeled trees with n nodes satisfies*

$$\frac{n^{n-2}}{n!} \le T_n \le 4^{n-1}.$$

The exact form of this lower bound does not matter much; we can conclude, just to have a statement simpler to remember, that the number of unlabeled trees on n nodes is larger than 2^n if n is large enough ($n > 30$ if you work it out). So we get, at least for $n > 30$, the following bounds, which are easy to remember:

$$2^n \le T_n \le 4^n.$$

The planar code is far from optimal; every unlabeled tree has many different codes (depending on how we draw it in the plane and how we choose the root), and not every 0–1 sequence of length $2(n-1)$ is a code of a tree (for example, it must start with a 1 and have the same number of 0's as 1's). Still, the planar code is quite an efficient way of encoding unlabeled trees: It uses less than $2n$ bits for trees with n nodes. Since there are more than 2^n unlabeled trees (at least for $n > 30$), we could not possibly get by with codes of length n: there are just not enough of them.

In contrast to what we know for labeled trees, we don't know a simple formula for the number of unlabeled trees on n nodes, and probably none exists. According to a difficult result of George Pólya, the number of unlabeled trees on n nodes is asymptotically $an^{-5/2}b^n$, where $a = 0.5349\ldots$ and $b = 2.9557\ldots$ are real numbers defined in a complicated way.

8.5.1 Does there exist an unlabeled tree with planar code
(a) 1111111100000000; (b) 1010101010101010; (c) 1100011100?

Review Exercises

8.5.2 Let G be a connected graph, and e an edge of G. Prove that e is not a cut-edge if and only if it is contained in a cycle of G.

8.5.3 Prove that a graph with n nodes and m edges has at least $n-m$ connected components.

8.5.4 Prove that if a tree has a node of degree d, then it has at least d leaves.

8.5.5 Find the number of unlabeled trees on 6 nodes.

8.5.6 A *double star* is a tree that has exactly two nodes that are not leaves. How many unlabeled double stars are there on n nodes?

8.5.7 Construct a tree from a path of length $n-3$ by creating two new nodes and connecting them to the same endpoint of the path. How many different labeled trees do you get from this tree?

8.5.8 Consider any table with 2 rows and $n - 1$ columns; the first row holds $1, 2, 3, \ldots, n - 1$; the second row holds arbitrary numbers between 1 and n. Construct a graph on nodes labeled $1, \ldots, n$ by connecting the two nodes in each column of our table.

(a) Show by an example that this graph is not always a tree.

(b) Prove that if the graph is connected, then it is a tree.

(c) Prove that every connected component of this graph contains at most one cycle.

8.5.9 Prove that in every tree, any two paths with maximum length have a node in common. This is not true if we consider two maximal (i.e., nonextendable) paths.

8.5.10 If C is a cycle, and e is an edge connecting two nonadjacent nodes of C, then we call e a *chord* of C. Prove that if every node of a graph G has degree at least 3, then G contains a cycle with a chord.

[Hint: Follow the proof of the theorem that a tree has a node of degree 1.]

8.5.11 Take an n-cycle, and connect two if its nodes at distance 2 by an edge. Find the number of spanning trees in this graph.

8.5.12 A (k, l)-*dumbbell graph* is obtained by taking a complete graph on k (labeled) nodes and a complete graph on l (labeled) nodes, and connecting them by a single edge. Find the number of spanning trees of a dumbbell graph.

8.5.13 Prove that if $n \geq 2$, then in Theorem 8.5.1 both inequalities are strict.

9
Finding the Optimum

9.1 Finding the Best Tree

A country with n towns wants to construct a new telephone network to connect all towns. Of course, they don't have to build a separate line between every pair of towns; but they do need to build a connected network; in our terms, this means that the graph of direct connections must form a connected graph. Let's assume that they don't want to build a direct line between towns that can be reached otherwise (there may be good reasons for doing so, as we shall see later, but at the moment let's assume their only goal is to get a connected network). Thus they want to build a minimal connected graph with these nodes, i.e., a tree.

We know that no matter which tree they choose to build, they have to construct $n-1$ lines. Does this mean that it does not matter which tree they build? No, because lines are not equally easy to build. Some lines between towns may cost much more than some other lines, depending on how far the towns are, whether there are mountains or lakes between them, etc. So the task is to find a spanning tree whose total cost (the sum of costs of its edges) is minimal.

How do we know what these costs are? Well, this is not something mathematics can tell you; it is the job of the engineers and economists to estimate the cost of each possible line in advance. So we just assume that these costs are given.

At this point, the task seems trivial (very easy) again: Just compute the cost of each tree on these nodes, and select the tree with smallest cost.

We dispute the claim that this is easy. The number of trees to consider is enormous: We know by Cayley's Theorem (Theorem 8.3.1) that the number of labeled trees on n nodes is n^{n-2}. So for 10 cities, we'd have to look at 10^8 (one hundred million) possible trees; for 20 cities, the number is astronomical (more than 10^{20}). We have to find a better way to select an optimal tree; and that's the point where mathematics comes to the rescue.

There is this story about the pessimist and the optimist: They each get a box of assorted candies. The optimist always picks the best; the pessimist always eats the worst (to save the better candies for later). So the optimist always eats the best available candy, and the pessimist always eats the worst available candy; and yet, they end up with eating same candies.

So let's see how the optimistic government would build the telephone network. They start with raising money; as soon as they have enough money to build a line (the cheapest line), they build it. Then they wait until they have enough money to build a second connection. Then they wait until they have enough money to build a third connection... It may happen that the third-cheapest connection forms a triangle with the first two (say, three towns are close to each other). Then, of course, they skip this and raise enough money to build the fourth-cheapest connection.

At any time, the optimistic government will wait until they have enough money to build a connection between two towns that are not yet connected by any path, and build this connection.

Finally, they will get a connected graph on the n nodes representing the towns. The graph does not contain a cycle, since the edge of the cycle constructed last would connect two towns that are already accessible from each other through the other edges of the cycle. So, the graph they get is indeed a tree.

But is this network the least expensive possible? Could the cheap attitude at the beginning backfire and force the government to spend much more at the end? We'll prove below that our optimistic government has undeserved success: The tree they build is as inexpensive as possible.

Before we jump into the proof, we should discuss why we said that the government's success was "undeserved." We show that if we modify the task a little, the same optimistic approach might lead to very bad results.

Let us assume that for reasons of reliability, they require that between any two towns there should be at least two paths with no edge in common (this guarantees that when a line is inoperational because of failure or maintenance, any two towns can still be connected). For this, $n-1$ lines are not enough ($n-1$ edges forming a connected graph must form a tree, but then if any edge is deleted, the rest will not be connected any more). But n lines suffice: All we have to do is to draw a single cycle through all the towns. This leads to the following task:

> *Find a cycle with n given towns as nodes such that the total cost of constructing its edges is minimum.*

(This problem is one of the most famous tasks in mathematical optimization: it is called the *Traveling Salesman Problem*. We'll say more about it later.)

Our optimistic government would do the following: Build the cheapest line, then the second cheapest, then the third cheapest, etc., skipping the construction of lines that are superfluous: It will not build a third edge out of a town that already has two, and will not build an edge completing a cycle unless this cycle connects all nodes. Eventually, they get a cycle through all towns, but this is not necessarily the best! Figure 9.1 shows an example where the optimistic method (called "greedy" in this area of applied mathematics) gives a cycle that is quite a bit worse than optimal.

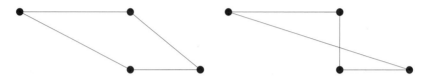

FIGURE 9.1. Failure of the greedy method. Construction costs are proportional to the distance. The first figure shows a cheapest (shortest) cycle through all 4 towns; the second shows the cycle obtained by the optimistic (greedy) method.

So the greedy method can be bad for the solution of a problem that is only slightly different from the problem of finding the cheapest tree. Thus the fact (to be proved below) that the optimistic government builds the best tree is indeed undeserved luck.

So let us return to the solution of the problem of finding a tree with minimum cost, and prove that the optimistic method yields a cheapest tree. The optimistic method is often called the *greedy Algorithm*; in the context of spanning trees, it is called *Kruskal's Algorithm*. Let us call the tree obtained by the greedy method the *greedy tree*, and denote it by F. In other words, we want to prove that any other tree would cost at least as much as the greedy tree (and so no one could accuse the government of wasting money and justify the accusation by exhibiting another tree that would have been cheaper).

So let G be any tree different from the greedy tree F. Let us imagine the process of constructing F, and the step when we first pick an edge that is not an edge of G. Let e be this edge. If we add e to G, we get a cycle C. This cycle is not fully contained in F, so it has an edge f that is not an edge of F (Figure 9.2). If we add the edge e to G and then delete f, we get a (third) tree H. (Why is H a tree? Give an argument!)

We want to show that H *is at most as expensive as* G. This clearly means that e is at most as expensive as f. Suppose (by indirect argument) that f is cheaper than e.

Now comes a crucial question: Why didn't the optimistic government select f instead of e at this point in time? The only reason could be that

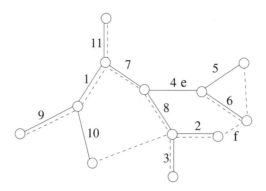

FIGURE 9.2. The greedy tree is optimal

f was ruled out because it would have formed a cycle C' with the edges of F already selected. But all these previously selected edges are edges of G, since we are inspecting the step when the first edge not in G was added to F. Since f itself is an edge of G, it follows that all edges of C' are edges of G, which is impossible, since G is a tree. This contradiction proves that f cannot be cheaper than e and hence G cannot be cheaper than H.

So we replace G by this tree H that is not more expensive. In addition, the new tree H has the advantage that it coincides with F in more edges, since we deleted from G an edge not in F and added an edge in F. This implies that if H is different from F and we repeat the same argument again and again, we get trees that are not more expensive than G, and coincide with F in more and more edges. Sooner of later we must end up with F itself, proving that F was no more expensive than G.

9.1.1 A pessimistic government could follow the following logic: If we are not careful, we may end up with having to build that extremely expensive connection through the mountain; so let us decide right away that building this connection is not an option, and mark it as "impossible." Similarly, let us find the second most expensive line and mark it "impossible," etc. Well, we cannot go on like this forever: We have to look at the graph formed by those edges that are still possible, and this "possibility graph" must stay connected. In other words, if deleting the most expensive edge that is still possible from the possibility graph would destroy the connectivity of this graph, then like it or not, we have to build this line. So we build this line (the pessimistic government ends up building the most expensive line among those that are still possible). Then they go on to find the most expensive line among those that are still possible and not yet built, mark it impossible if this does not disconnect the possibility graph, etc.

Prove that the pessimistic government will have the same total cost as the optimistic.

9.1.2 Formulate how the pessimistic government will construct a cycle through all towns. Show by an example that they don't always get the cheapest solution.

9.2   The Traveling Salesman Problem

Let us return to the question of finding a cheapest possible cycle through all the given towns: We have n towns (points) in the plane, and for any two of them we are given the "cost" of connecting them directly. We have to find a cycle with these nodes such that the cost of the cycle (the sum of the costs of its edges) is as small as possible.

This problem is one of the most important in the area of *combinatorial optimization*, the field dealing with finding the best possible design in various combinatorial situations, like finding the optimal tree discussed in the previous section. It is called the *Traveling Salesman Problem*, and it appears in many disguises. Its name comes from the version of the problem where a traveling salesman has to visit all towns in a region and then return to his home, and of course, he wants to minimize his travel costs. It is clear that mathematically, this is the same problem. It is easy to imagine that one and the same mathematical problem appears in connection with designing optimal delivery routes for mail, optimal routes for garbage collection, etc.

The following important question leads to the same mathematical problem, except on an entirely different scale. A machine has to drill a number of holes in a printed circuit board (this number could be in the thousands), and then return to the starting point. In this case, the important quantity is the *time* it takes to move the drilling head from one hole to the next, since the total time a given board has to spend on the machine determines the number of boards that can be processed in a day. So if we take the time needed to move the head from one hole to another as the "cost" of this edge, we need to find a cycle with the holes as nodes, and with minimum cost.

The Traveling Salesman Problem is closely related to Hamiltonian cycles. First of all, a traveling salesman tour is just a Hamiltonian cycle in the complete graph on the given set of nodes. But there is a more interesting connection: *The problem of whether a given graph has a Hamiltonian cycle can be reduced to the Traveling Salesman Problem.*

Let G be a graph with n nodes. We define the "distance" of two nodes as follows: adjacent nodes have distance 1; nonadjacent nodes have distance 2.

What do we know about the Traveling Salesman Problem on the set of nodes of G with this new distance function? If the graph contains a Hamiltonian cycle, then this is a traveling salesman tour of length n. If the graph has no Hamiltonian cycle, then the shortest traveling salesman tour has length at least $n + 1$. This shows that any algorithm that solves the Traveling Salesman Problem can be used to decide whether or not a given graph has a Hamiltonian cycle.

The Traveling Salesman Problem is much more difficult than the problem of finding the cheapest tree, and there is no algorithm to solve it that would be anywhere nearly as simple, elegant and efficient as the "optimistic" algorithm discussed in the previous section. There are methods that work quite well most of the time, but they are beyond the scope of this book.

But we want to show at least one simple algorithm that, even though it does not give the best solution, never loses more than a factor of 2. We describe this algorithm in the case where the cost of an edge is just its length, but it would not make any difference to consider any other measure (like time, or the price of a ticket), as long as the costs $c(ij)$ satisfy the *triangle inequality*:

$$c(ij) + c(jk) \geq c(ik). \tag{9.1}$$

(Distances in Euclidean geometry satisfy this condition by classical results in geometry: The shortest route between two points is a straight line. Airfares sometimes don't satisfy this inequality: It may be cheaper to fly from New York to Chicago to Philadelphia then to fly from New York to Philadelphia. But in this case, of course, we might consider the flight New York–Chicago–Philadelphia as one "edge," which does not count as a visit in Chicago. The distance function on a graph we introduced above when we discussed the connection between the Traveling Salesman Problem and Hamiltonian cycles satisfies the triangle inequality.)

We begin by solving a problem we know how to solve: Find a cheapest tree connecting up the given nodes. We can use any of the algorithms discussed in the previous section for this. So we find the cheapest tree T, with total cost c.

Now, how does this tree help in finding a tour? One thing we can do is to walk around the tree just as we did when constructing the "planar code" of a tree in the proof of Theorem 8.5.1 (see Figure 8.6). This certainly gives a walk that goes through each town at least once, and returns to the starting point.

Of course, this walk may pass through some of the towns more than once. But this is good for us: We can make shortcuts. If the walk takes us from i to j to k, and we have seen j already, we can proceed directly from i to k. Doing such shortcuts as long as we can, we end up with a tour that goes through every town exactly once (Figure 9.3). Let us call the algorithm described above the *Tree Shortcut Algorithm*.

Theorem 9.2.1 *If the costs in a Traveling Salesman Problem satisfy the triangle inequality (9.1), then the Tree Shortcut Algorithm finds a tour that costs less than twice as much as the optimum tour.*

Proof. The cost of the walk around the tree is exactly twice the cost c of T, since we used every edge twice. The triangle inequality guarantees that we have only shortened our walk by doing shortcuts, so the cost of the tour we found is not more than twice the cost of the cheapest spanning tree.

But we want to relate the cost of the tour we obtained to the cost of the optimum tour, not to the cost of the optimum spanning tree! Well, this is

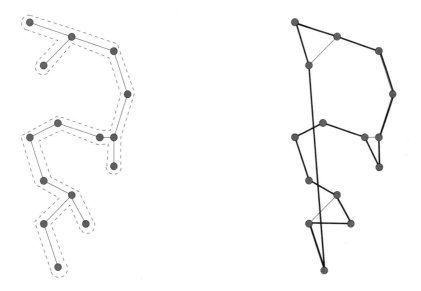

FIGURE 9.3. The cheapest tree connecting 15 given towns, the walk around it, and the tour produced by shortcuts (light edges on the right hand side are edges of the tree that are not used in the tour). Costs are proportional to distances.

easy now: *The cost of a cheapest spanning tree is always less than the cost of the cheapest tour.* Why? Because we can omit any edge of the cheapest tour to get a spanning tree. This is a very special kind of tree (a path), and as a spanning tree it may or may not be optimal. However, its cost is certainly not smaller than the cost of the cheapest tree, but smaller than the cost of the optimal tour, which proves the assertion above.

To sum up, the cost of the tour we constructed is at most twice that of the cheapest spanning tree, which in turn is less than twice the cost of a cheapest tour.   □

9.2.1 Is the tour in Figure 9.3 shortest possible?

9.2.2 Prove that if all costs are proportional to distances, then a shortest tour cannot intersect itself.

Review Exercises

9.2.3 Prove that if all edge-costs are different, then there is only one cheapest tree.

9.2.4 Describe how you can find a spanning tree for which (a) the product of the edge-costs is minimal; (b) the maximum of the edge-costs is minimal.

9.2.5 In a real-life government, optimists and pessimists win in unpredictable order. This means that sometimes they build the cheapest line that does not create a cycle with those lines already constructed; sometimes they mark the most expensive lines "impossible" until they get to a line that cannot be marked impossible without disconnecting the network, and then they build it. Prove that they still end up with the same cost.

9.2.6 If the seat of the government is town r, then quite likely the first line constructed will be the cheapest line out of r (to some town s, say), then the cheapest line that connects either r or s to a new town, etc. In general, there will be a connected graph of telephone lines constructed on a subset S of the towns including the capital, and the next line will be the cheapest among all lines that connect S to a node outside S. Prove that the lucky government still obtains a cheapest possible tree.

9.2.7 Find the shortest tour through the points of a (a) 3×3 square grid; (b) 4×4 square grid; (c) 5×5 square grid; (d) generalize to $n \times m$ grids.

9.2.8 Show by an example that if we don't assume the triangle inequality, then the tour found by the Tree Shortcut Algorithm can be longer than 1000 times the optimum tour.

10
Matchings in Graphs

10.1 A Dancing Problem

At the prom, 300 students took part. They did not all know each other; in fact, every girl knew exactly 50 boys and every boy knew exactly 50 girls (we assume, as before, that acquaintance is mutual).

We claim that the students can all dance simultaneously so that only pairs who know each other dance with each other.

Since we are talking about acquaintances, it is natural to describe the situation by a graph (or at least, imagine the graph that describes it). So we draw 300 nodes, each representing a student, and connect two of them if they know each other. Actually, we can make the graph a little simpler: the fact that two boys, or two girls, know each other plays no role whatsoever in this problem: so we don't have to draw those edges that correspond to such acquaintances. We can then arrange the nodes, conveniently, so that the nodes representing boys are on the left, and nodes representing girls are on the right; then every edge will connect a node on the left to a node on the right. We shall denote the set of nodes on the left by A, the set of nodes on the right by B.

This way we obtain a special kind of graph, called a *bipartite graph*. Figure 10.1 shows such a graph (of course, depicting a smaller party). The thick edges show one way to pair up people for dancing. Such a set of edges is called a *perfect matching*.

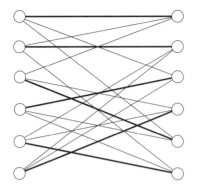

FIGURE 10.1. A bipartite graph with a perfect matching.

To be precise, let's give the definitions of these terms: A graph is *bipartite* if its nodes can be partitioned into two classes, say A and B, such that every edge connects a node in A to a node in B. A *perfect matching* is a set of edges such that every node is incident with exactly one of these edges.

After this, we can formulate our problem in the language of graph theory as follows: We have a bipartite graph with 300 nodes, in which every node has degree 50. We want to prove that it contains a perfect matching.

As before, it is good idea to generalize the assertion to any number of nodes. Let's be daring and guess that the numbers 300 and 50 play no role whatsoever. The only condition that matters is that all nodes have the same degree (and this is not 0). Thus we set out to prove the following theorem, named after the Hungarian mathematician D. König (who wrote the first book on graph theory).

Theorem 10.1.1 *If every node of a bipartite graph has the same degree $d \geq 1$, then it contains a perfect matching.*

Before proving the theorem, it will be useful to solve some exercises, and then discuss another problem in the next section.

10.1.1 It is obvious that for a bipartite graph to contain a perfect matching, it is necessary that $|A| = |B|$. Show that if every node has the same degree, then this is indeed so.

10.1.2 Show by examples that the conditions formulated in the theorem cannot be dropped:

(a) A nonbipartite graph in which every node has the same degree need not contain a perfect matching.

(b) A bipartite graph in which every node has positive degree (but not all the same) need not contain a perfect matching.

10.1.3 Prove Theorem 10.1.1 for $d = 1$ and $d = 2$.

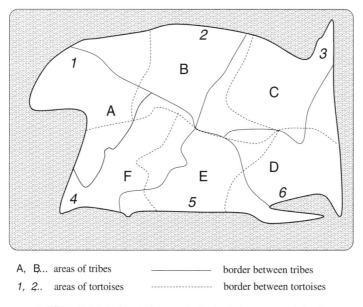

A, B... areas of tribes   ——————   border between tribes

1, 2.. areas of tortoises   ·····················   border between tortoises

FIGURE 10.2. Six tribes and six tortoises on an island

10.2   Another matching problem

An island is inhabited by six tribes. They are on good terms and split up the island between them, so that each tribe has a hunting territory of 100 square miles. The whole island has an area of 600 square miles.

The tribes decide that they all should choose new totems. They decide that each tribe should pick one of the six species of tortoise that live on the island. Of course, they want to pick different totems, and in such a way that the totem of each tribe should occur somewhere on their territory.

It is given that the territories where the different species of tortoises live don't overlap, and they have the same area, 100 square miles (so it also follows that every part of the island is inhabited by some kind of tortoise). Of course, the way the tortoises divide up the island may be entirely different from the way the tribes do (Figure 10.2)

We want to prove that such a selection of totems is always possible.

To see the significance of the conditions, let's assume that we did not stipulate that the area of each tortoise species is the same. Then some species could occupy more, say, 200 square miles. But then it could happen that two of tribes are living on exactly these 200 square miles, and so their only possible choice for a totem would be one and the same species.

Let's try to illustrate our problem by a graph. We can represent each tribe by a node, and also each species of tortoise by a node. Let us connect a tribe-node to a tortoise-node if the species occurs somewhere on the territory of the tribe (we could also say that the tribe occurs on the territory of the species, just in case the tortoises want to pick totems too). Drawing the tribe-nodes on the left and the tortoise-nodes on the right makes it clear that we get a bipartite graph (Figure 10.3). And what is it that we want to prove? It is that this graph has a perfect matching!

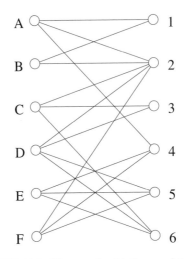

FIGURE 10.3. The graph of tribes and tortoises

So this is very similar to the problem discussed (but not solved!) in the previous section: We want to prove that a certain bipartite graph has a perfect matching. Theorem 10.1.1 says that for this conclusion it suffices to know that every node has the same degree. But this is too strong a condition; it is not at all fulfilled in our example (tribe B has only two tortoises to choose from, while tribe D has four).

So what property of this graph should guarantee that a perfect matching exists? Turning this question around: What would *exclude* a perfect matching?

For example, it would be bad if a tribe could not find any tortoises on its own territory. In the graph, this would correspond to a node with degree 0. Now this is not a danger, since we know that tortoises occur everywhere on the island.

It would also be bad (and this has come up already) if two tribes could only choose one and the same tortoise. But then this tortoise would have an area of at least 200 square miles, which is not the case. A somewhat more subtle sort of trouble would arise if three tribes had only two tortoises on their combined territory. But this, too, is impossible: The two species of tortoises would cover an area of at least 300 square miles, so one of them would have to cover more than 100. More generally, we can see that the combined territory of any k tribes holds at least k species of tortoises. In terms of the graph, this means that for any k nodes on the left, there are at least k nodes on the right connected to at least one of them. We'll see in the next section that this is all we need to observe about this graph.

10.3   The Main Theorem

Now we state and prove a fundamental theorem about perfect matchings. This will complete the solution of the problem about tribes and tortoises, and (with some additional work) of the problem about dancing at the prom (and some problems further down the road from the prom, as its name shows).

Theorem 10.3.1 (The Marriage Theorem) *A bipartite graph has a perfect matching if and only if $|A| = |B|$ and for any subset of (say) k nodes of A there are at least k nodes of B that are connected to at least one of them.*

This important theorem has many variations; some of these occur in the exercises. These were discovered by the German mathematician G. Frobenius, by the Hungarian D. König, the American P. Hall, and others.

Before proving this theorem, let us discuss one more question. If we interchange "left" and "right," perfect matchings remain perfect matchings. But what happens to the condition stated in the theorem? It is easy to see that it remains valid (as it should). To see this, we have to argue that if we pick any set S of k nodes in B, then they are connected to at least k nodes in A. Let $n = |A| = |B|$ and let us color the nodes in A connected to nodes in S black, the other nodes white (Figure 10.4). Then the white nodes are connected to at most $n - k$ nodes (since they are not connected to any node in S). Since the condition holds "from left to right," the number of white nodes is at most $n - k$. But then the number of black nodes is at least k, which proves that the condition also holds "from right to left."

Proof. Now we can turn to the proof of Theorem 10.3.1. We shall have to refer to the condition given in the theorem so often that it will be convenient to call graphs satisfying this conditions "good" (just for the duration of this proof). Thus a bipartite graph is "good" if it has the same number of nodes left and right, and any k "left" nodes are connected to at least k "right" nodes.

It is obvious that every graph with a perfect matching is "good," so what we need to prove is the converse: *Every "good" graph contains a perfect matching.* For a graph on just two nodes, being "good" means that these two nodes are connected. Thus for a graph to have a perfect matching means that it can be partitioned into "good" graphs with 2 nodes. (To partition a graph means that we divide the nodes into classes, and keep an edge between two nodes only if they are in the same class.)

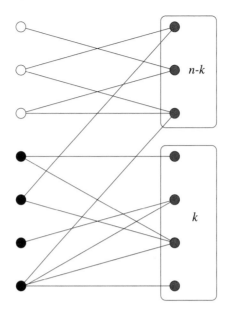

FIGURE 10.4. The good graph is also good from right to left.

Now our plan is to partition our graph into two "good" parts, then partition each of these into two "good" parts, etc., until we get "good" parts with 2 nodes. Then the edges that remain form a perfect matching. To carry out this plan, it suffices to prove that

if a "good" bipartite graph has more than 2 nodes, then it can be partitioned into two good bipartite graphs.

Let us try a very simple partition first: Select nodes $a \in A$ and $b \in B$ that are connected by an edge; let these two nodes be the first part, and the remaining nodes the other. There is no problem with the first part: it is "good." But the second part may not be good: It can have some set S of k nodes on the left connected to fewer than k nodes on the right (Figure 10.5). In the original graph, these k nodes were connected to at least k nodes in B; this can hold only if the kth such node was the node b. Let T denote the set of neighbors of S in the original graph. What is important to remember is that $|S| = |T|$.

Now we try another way of partitioning the graph: We take $S \cup T$ (together with the edges between them) as one part, and the rest of the nodes as the other. (This rest is not empty: The node a belongs to it, for example.)

Let's argue that both these parts are "good." Take the first graph first. Take any subset of, say, j nodes in S (the left-hand side of the first graph). Since the original graph was good, they are connected to at least j nodes, which are all in T by the definition of T.

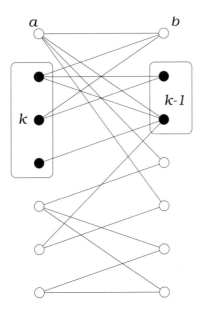

FIGURE 10.5. Goodness lost when two nodes are removed

For the second graph, it follows similarly that it is good if we interchange "left" and "right." This completes the proof. □

We still have to prove Theorem 10.1.1. This is now quite easy and is left to the reader as Exercise 10.3.1.

10.3.1 Prove that if in a bipartite graph every node has the same degree $d \neq 0$, then the bipartite graph is "good" (and hence contains a perfect matching; this proves Theorem 10.1.1).

10.3.2 Suppose that in a bipartite graph, for any subset X of nodes of A there are at least $|X|$ nodes in B that are connected to one of them (but in contrast to Theorem 10.3.1, we don't assume that $|A| = |B|$). Prove that there is a set of edges that match every node of A with a node of B, where different nodes of A are matched with different nodes of B (but some nodes of B may remain unmatched).

10.4   How to Find a Perfect Matching

We have a condition for the existence of a perfect matching in a graph that is *necessary and sufficient*. Does this condition settle this issue once and for all? To be more precise: Suppose that somebody gives us a bipartite graph; what is a good way to *decide* whether it contains a perfect matching? And how do we *find* a perfect matching if there is one?

We may assume that $|A| = |B|$ (where, as before, A is the set of nodes on the left and B is the set of nodes on the right). This is easy to check, and if it fails, then it is obvious that no perfect matching exists, and we have nothing else to do.

One thing we can try is to look at all subsets of the edges, and see whether any of these is a perfect matching. It is easy enough to do so; but there are terribly many subsets to check! Say, in our introductory example, we have 300 nodes, so $|A| = |B| = 150$; every node has degree 50, so the number of edges is $150 \cdot 50 = 7500$; the number of subsets of a set of this size is $2^{7500} > 10^{2257}$, a number that is more than astronomical.

We can do a little bit better if instead of checking all subsets of the edges, we look at all possible ways to pair up elements of A with elements of B, and check whether any of these pairings matches only nodes that are connected to each other by an edge. Now the number of ways to pair up the nodes is "only" $150! \approx 10^{263}$. Still hopeless.

Can we use Theorem 10.3.1? To check that the necessary and sufficient condition for the existence of a perfect matching is satisfied, we have to look at every subset S of A, and see whether the number of it neighbors in B is at least as large as S itself. Since the set A has $2^{150} \approx 10^{45}$ subsets, this takes a much smaller number of cases to check than any of the previous possibilities, but still astronomical!

So Theorem 10.3.1 does not really help too much in deciding whether a given graph has a perfect matching. We have seen that it does help in *proving* that certain properties of a graph imply that the graph has a perfect matching. We'll come back to this theorem later and discuss its significance. Right now, we have to find some other way to deal with our problem.

Let us introduce one more expression: By a *matching* we mean a set of edges that have no endpoint in common. A perfect matching is the special case when, in addition, the edges cover all the nodes. But a matching can be much smaller: the empty set, or any edge by itself, is a matching.

Let's try to construct a perfect matching in our graph by starting with the empty set and building up a matching one by one. So we select two nodes that are connected, and mark the edge between them; then we select two other nodes that are connected, and mark the edge between them etc. we can do this until no two unmatched nodes are connected by an edge. The edges we have marked form a matching M. This is often called the *greedy matching*, since it is constructed greedily, without consideration for the future consequences of our choice. If we are lucky, then the greedy matching is perfect, and we have nothing else to do. But what do we do if M is not perfect? Can we conclude that the graph has no perfect matching at all? No, we cannot; it may happen that the graph has a perfect matching, but we made some unlucky choices when selecting the edges of M.

10.4.1 Show by an example that it may happen that a bipartite graph G has a perfect matching, but if we are unlucky, the greedy matching M constructed above is not perfect.

10.4.2 Prove that if G has a perfect matching, then every greedy matching matches up at least half of the nodes.

So suppose that we have constructed a matching M that is not perfect. We have to try to increase its size by "backtracking," i.e., by deleting some of its edges and replacing them by more edges. But how do we find the edges we want to replace?

The trick is the following. We look for a path P in G of the following type: P starts and ends at nodes u and v that are unmatched by M; and every second edge of P belongs to M (Figure 10.6). Such a path is called an *augmenting path*. It is clear that an augmenting path P contains an odd number of edges, and in fact, the number of its edges not in M is one larger than the number of its edges in M.

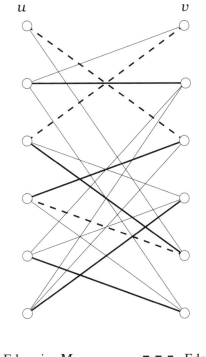

——— Edges in M - - - Edges in P
                                    not in M

FIGURE 10.6. An augmenting path in a bipartite graph.

If we find an augmenting path P, we can delete those edges of P that are in M and replace them by those edges of P that are not in M. It is clear that this results in a matching M' that is larger than M by one edge. (The fact that M' is a matching follows from the observation that the remaining edges of M cannot contain any node of P: The two endpoints of P were supposed to be unmatched, while the interior nodes of P were matched by edges of M that we deleted.) So we can repeat this until we get either a perfect matching or a matching M for which no augmenting path exists.

So we have two questions to answer: how do we find an augmenting path if it exists? And if it does not exist, does this mean that there is no perfect matching at all? It will turn out that an answer to the first question will also imply the (affirmative) answer to the second.

Let U be the set of unmatched nodes in A and let W be the set of unmatched nodes in B. As we noted, any augmenting path must have an odd number of edges, and hence it must connect a node in U to a node in W. Let us try to find such an augmenting path starting from some node in U. Let's say that a path Q is *almost augmenting* if it starts at a node in U, ends at a node in A, and every second edge of it belongs to M. An almost augmenting path must have an even number of edges, and must end with an edge of M.

What we want to do is to find the set of nodes in A that can be reached on an almost augmenting path. Let's agree that we consider a node in U to be an almost augmenting path in itself (of length 0); then we know that every node in U has this property. Starting with $S = U$, we build up a set S gradually. At any stage, the set S will consist of nodes we already know are reachable by some almost augmenting path. We denote by T the set of nodes in B that are matched with nodes in S (Figure 10.7). Since the nodes of U have nothing matched with them and they are all in S, we have

$$|S| = |T| + |U|.$$

We look for an edge that connects a node $s \in S$ to some node $r \in B$ that is *not* in T. Let Q be an almost augmenting path starting at some node $u \in U$ and ending at s. Now there are two cases to consider:

— If r is unmatched (which means that it belongs to W), then by appending the edge sr to Q we get an augmenting path P. So we can increase the size of M (and forget about S and T).

— If r is matched with a node $q \in A$, then we can append the edges sr and rq to Q to get an almost augmenting path from U to q. So we can add q to S.

So if we find an edge connecting a node in S to a node not in T, we can increase either the size of M or the set S (and leave M as it was). Sooner or later we must encounter a situation where either M is a perfect matching

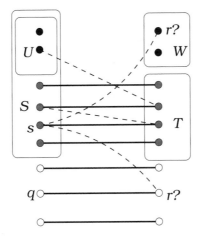

FIGURE 10.7. Reaching nodes by almost augmenting paths. Only edges on these paths, and of M, are shown.

(and we are done), or M is not perfect, but no edge connects S to any node outside T.

So what are we to do in this case? Nothing! If this occurs, we can conclude that there is no perfect matching at all. In fact, all neighbors of the set S are in T, and $|T| = |S| - |U| < |S|$. We know that this implies that there is no perfect matching at all in the graph.

Figure 10.8 shows how this algorithm finds a matching in the bipartite graph that is a subgraph of the "grid."

To sum up, we do the following. At any point in time, we will have a matching M and a set S of nodes in A that we know can be reached on almost augmenting paths. If we find an edge connecting S to a node not matched with any node in S, we can either increase the size of M or the set S, and repeat. If no such edge exists, then either M is perfect or no perfect matching exists at all.

Remark. In this chapter we restricted our attention to matchings in bipartite graphs. One can, of course, define matchings in general (nonbipartite) graphs. It turns out that both the necessary and sufficient condition given in Theorem 10.3.1 and the algorithm described in this section can be extended to nonbipartite graphs. However, this requires methods that are quite a bit more involved, which lie beyond the scope of this book.

10.4.3 Follow how the algorithm works on the graph in Figure 10.9.

10.4.4 Show how the description of the algorithm above contains a new proof of Theorem 10.3.1.

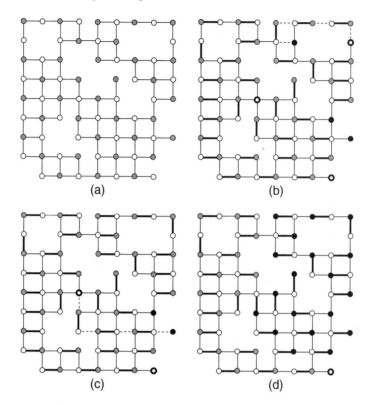

FIGURE 10.8. (a) The graph in which we want to find a perfect matching. (b) Pick a starting matching, and mark the unmatched nodes. There are 3 black and 3 white unmatched nodes. Broken lines indicate an augmenting path. (c) The new matching and the unmatched nodes after augmentation. Broken lines indicate a new augmenting path (much longer this time). (d) The final situation: Nodes that are accessible on almost augmenting paths are marked black. They have fewer neighbors than their number, so the matching is maximum.

Review Exercises

10.4.5 Is there a bipartite graph with degrees $3, 3, 3, 3, 3, 3, 3, 3, 3, 5, 6, 6$? (These can be distributed in the two classes of nodes arbitrarily.)

10.4.6 A bipartite graph has 16 nodes of degree 5, and some nodes of degree 8. We know that all degree-8 nodes are on the left hand side. How many degree 8 nodes can the graph have?

10.4.7 Let G be a bipartite graph with the same number of nodes on both sides. Suppose that every nonempty subset A on the left has at least $|A| + 1$ neighbors on the right. Prove that each edge of G can be extended to a perfect matching of G.

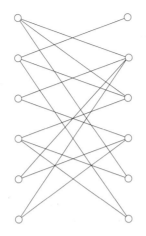

FIGURE 10.9. A graph for trying out the algorithm.

10.4.8 Now suppose that we have the weaker condition that every nonempty subset A on the left has at least $|A| - 1$ neighbors on the right. Prove that G contains a matching that matches up all but one node on each side.

10.4.9 Let G be a bipartite graph with m nodes on both sides. Prove that if each node has degree larger than $m/2$, then it has a perfect matching.

10.4.10 Does the graph in Figure 10.10 have a perfect matching?

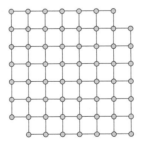

FIGURE 10.10. A truncated chessboard.

10.4.11 Draw a graph whose nodes are the subsets of $\{a, b, c\}$, and for which two nodes are adjacent if and only if they are subsets that differ in exactly one element.

(a) What is the number of edges and nodes in this graph? Can you name this graph?

(b) Is this graph connected? Does it have a perfect matching? Does it have a Hamilton cycle?

10.4.12 Draw a graph whose nodes are the 2-subsets of $\{a, b, c, d, e\}$, and two nodes are adjacent if and only if they are disjoint subsets.

(a) Show that you get the Petersen graph (Figure 7.13).

(b) How many perfect matchings does the Petersen graph have?

10.4.13 (a) How many perfect matchings does a path on n nodes have? (b) How many matchings (not necessarily perfect) does a path on n nodes have? [Find a recurrence first.] (b) How many matchings does a cycle on n nodes have?

10.4.14 Which 2-regular bipartite graph with n nodes has the largest number of perfect matchings?

10.4.15 How many perfect matchings does the "ladder" with $2n$ nodes (Figure 10.11) have?

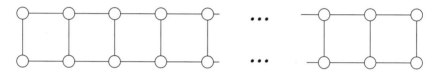

FIGURE 10.11. The ladder graph.

11

Combinatorics in Geometry

11.1 Intersections of Diagonals

At first you may be surprised: What is the connection between combinatorics and geometry? There are many geometric questions that can be solved by combinatorial methods, but the opposite case may also occur: We can solve combinatorial exercises and problems using geometrical tools.

Consider a convex polygon with n vertices. (We call a polygon *convex* if every angle of it is convex, i.e., less than $180°$.) Assume that it has no 3 diagonals going through the same point. How many intersection points do the diagonals have? (The vertices are not counted as intersections, and we do not consider intersections of diagonals outside the n-gon. In Figure 11.1, the black point is a "good" intersection, so it is counted, but the circled point outside the polygon is not counted.)

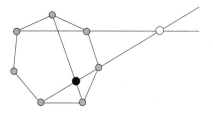

FIGURE 11.1.

The first natural idea to solve this problem is the following: Count the intersection points on each diagonal, and then add up these numbers. Let's follow through this method in the case of a hexagon (Figure 11.2). First consider a diagonal which connects two second-neighboring vertices of the hexagon, say these are A and C. This diagonal AC intersects all three diagonals that start in B, which means that there are three intersections on the diagonal AC. There are six diagonals of this type, so altogether we count $6 \cdot 3 = 18$ intersections.

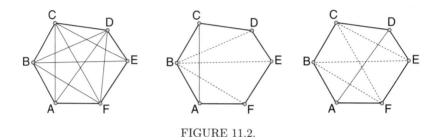

FIGURE 11.2.

Now consider a diagonal that connects two opposite vertices of the hexagon, say AD. One can see in the figure that there are four intersections on this diagonal. We have three diagonals of this type, which means that we get $3 \cdot 4 = 12$ further intersections.

Is it true that we have $18+12 = 30$ intersections? We must be more careful! The intersection of the diagonals AC and BD was counted twice: Once when we considered the intersections on AC, and also when we counted the intersections on BD. The same holds for any other intersection too: we counted all of them exactly twice. So we have to divide our result by 2, so our final, correct result is 15, which can be easily checked in the figure.

One can see that already for the case of such a small n, we have to consider several cases, so our method is too complicated, even though it can be carried through for arbitrary n (try it!). But we can find a much more elegant enumeration of the intersections if we use our combinatorial knowledge.

Label every intersection point by the endpoints of the diagonals intersecting at this point. For instance, the intersection of the diagonals AC and BD gets the label $ABCD$, the intersection of the diagonals AD and CE gets the label $ACDE$, and so on. Is this labeling good, by which we mean that different intersection points get different labels? The labeling is good, since the labels A, B, C, D are given only to the intersection of the diagonals of the convex quadrilateral $ABCD$ (AC and BD). Furthermore, every set of 4 vertices are used to label an intersection point of the diagonals; for instance, the 4-tuple $ACEF$ denotes the intersection of the diagonals AE and CF (Figure 11.3).

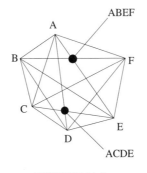

FIGURE 11.3.

Now if we want to count all the intersection points, it suffices to count quadruples of vertices; the number of intersections of the diagonals is just the number of 4-element subsets of the set of vertices. So if $n = 6$, then it is $\binom{6}{4}$, which we may compute even faster, if we recall that it is the same as $\binom{6}{2} = \frac{6 \cdot 5}{2 \cdot 1} = 15$. In general, we get $\binom{n}{4}$ intersections of diagonals of a convex n-gon.

11.1.1 How many diagonals does a convex n-gon have?

11.2   Counting regions

Let us draw n lines in the plane. These lines divide the plane into some number of regions. How many regions do we get?

The first thing to notice is that this question does not have a single answer. For example, if we draw two lines, we get 3 regions if the two are parallel, and 4 regions if they are not.

OK, let us assume that no two of the lines are parallel; then 2 lines always give us 4 regions. But if we go on to three lines, we get 6 regions if the lines go through one point, and 7 regions if they do not (Figure 11.4).

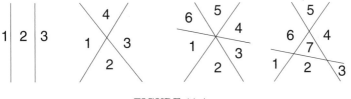

FIGURE 11.4.

OK, let us also exclude this, and assume that no 3 lines go through the same point. One might expect that the next unpleasant example comes with 4 lines, but if you experiment with drawing 4 lines in the plane, with no two parallel and no three going through the same point, then you invariably get 11 regions. In fact, we'll have a similar experience for any number of lines (check this in Figure 11.5).

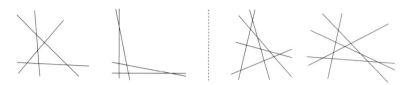

FIGURE 11.5. Four lines determine 11 regions, five lines determine 16.

A set of lines in the plane such that no two are parallel and no three go through the same point is said to be *in general position*. If we choose the lines "randomly," then accidents like two being parallel or three going through the same point will be very unlikely, so our assumption that the lines are in general position is quite natural.

Even if we accept that the number of regions is always the same for a given number of lines, the question still remains, what is this number? Let us collect our data in a little table (including also the observation that 0 lines divide the plane into 1 region, and 1 line divides the plane into 2):

0	1	2	3	4
1	2	4	7	11

Staring at this table for a while, we observe that each number in the second row is the sum of the number above it and the number to the left of it. This suggests a rule: The nth entry is n plus the previous entry. In other words: *If we have a set of $n - 1$ lines in the plane in general position, and add a new line (preserving general position), then the number of regions increases by n.*

Let us prove this assertion. How does the new line increase the number of regions? By cutting some of them into two. The number of additional regions is just the same as the number of regions intersected.

So, how many regions does the new line intersect? At a first glance, this is not easy to answer, since the new line can intersect very different sets of regions, depending on where we place it. But imagine walking along the new line, starting from very far away. We get to a new region every time we cross a line. So the number of regions the new line intersects is one larger than the number of crossing points on the new line with other lines.

Now, the new line crosses every other line (since no two lines are parallel), and it crosses them in different points (since no three lines go through the same point). Hence during our walk, we see $n - 1$ crossing points. So we see n different regions. This proves that our observation about the table is true for every n.

We are not done yet; what does this give for the number of regions? We start with 1 region for 0 lines, and then add to it $1, 2, 3, \ldots, n$. This way we get

$$1 + (1 + 2 + 3 + \cdots + n) = 1 + \frac{n(n+1)}{2}$$

(in the last step we used the "young Gauss' problem from Chapter 1). Thus we have proved:

Theorem 11.2.1 *A set of n lines in general position in the plane divides the plane into $1 + n(n+1)/2$ regions.*

11.2.1 Describe a proof of Theorem 11.2.1 using induction on the number of lines.

Let us give another proof of Theorem 11.2.1.

Proof. This time, we will not use induction, but rather try to relate the number of regions to other combinatorial problems. One gets a hint from writing the number in the form $1 + n + \binom{n}{2}$.

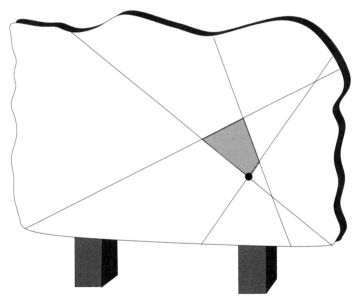

FIGURE 11.6.

Assume that the lines are drawn on a vertical blackboard (Figure 11.6), which is large enough so that all the intersection points appear on it. We also assume that no line is perfectly horizontal (it it is, we tilt the picture a little), and that the blackboard is very long, so that every line intersects the bottom edge of the blackboard. We may also assume that the lower edge of the blackboard tilts a little to the left (else, we tilt the blackboard by a tiny amount).

Now consider the lowest point in each region on the blackboard. Each region has exactly one lowest point, since all regions are finite, and the bordering lines are not horizontal. This lowest point is then an intersection point of two of our lines, or the intersection point of a line with the lower edge of the blackboard, or the lower left corner of the blackboard. Furthermore, each of these points is the lowest point of one and only one region. For example, if we consider any intersection point of two lines, then we see that four regions meet at this point, and the point is the lowest point of exactly one of them.

Thus the number of lowest points is the same as the number of intersection points of the lines, plus the number of intersection points between lines and the lower edge of the blackboard, plus one for the lower left corner of the blackboard. Since any two lines intersect, and these intersection points are all different (this is where we use that the lines are in general position), the number of such lowest points is $\binom{n}{2} + n + 1$. $\qquad\qquad\square$

11.3   Convex Polygons

Let us finish our excursion to combinatorial geometry with a problem that can be easily stated, but which is still unsolved. It is often called the *Happy End Problem*, because the mathematician who raised it, Esther Klein, and the mathematician who proved the first bounds on it, György Szekeres (together with Pál Erdős), got married shortly after.

We start with an exercise:

> Prove that if we are given five points in the plane such that no three of them are on a line, then we can always find four points among them that form the vertices of a convex quadrilateral.

For a quadrilateral, convexity means that the two diagonals intersect inside the quadrilateral. In Figure 11.7, the first quadrilateral is convex; the second is concave.

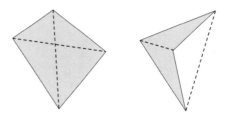

FIGURE 11.7. A convex and a concave quadrilateral.

In the sequel, for the sake of brevity we will say that certain points are in *general position*, if no three of them are on a line.

Imagine that our five points (in general position) are drawn on the black-board. Let us hammer nails into the points (only in imagination!) and stretch a rubber band around them. The rubber band encloses a convex polygon, the vertices of which are among the original points, and the other points are inside of this polygon. This polygon is called the *convex hull* of the points (Figure 11.8).

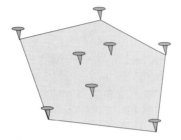

FIGURE 11.8. The convex hull of a set of points in the plane.

Using this construction, it is quite easy to solve our problem. What can we get as the convex hull of our five points? If this is a pentagon, then any four of these points form a convex quadrilateral, and we are done. If the convex hull is a quadrilateral, then this quadrilateral itself is the desired convex quadrilateral. So we may assume that the convex hull is a triangle. Let's denote the vertices of this triangle by A, B, and C, and the two other given points by D and E. The points D and E are inside the triangle ABC, so the line through D and E intersects the circumference of the triangle in two points. Assume that these two points are on the sides AB and AC (if this is not the case, then we can relabel the vertices of the triangle). Now you can easily show that the points B, C, D, and E form the vertices of a convex quadrilateral (see Figure 11.9).

The next fact in the line concerns convex pentagons.

> *If we have 9 points in the plane in general position, then one can choose five points among them that are the vertices of a convex pentagon.*

The proof of this statement is too lengthy to give it as an exercise, but the reader may try to work out a proof.

11.3.1 Draw eight points in the plane in such a way that no five of them span a convex pentagon.

One can give 16 points in general position in the plane such that no 6 of them form a convex hexagon. Nobody has been able to construct 17

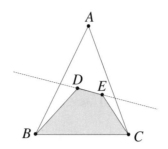

FIGURE 11.9. Finding a convex quadrilateral.

points with this property, and in fact (as we are writing this), substantial computer assisted research is underway to show that one can always find a convex hexagon among any 17 points in general position in the plane. More generally, we can ask, How many points (in general position) do we need to make sure that we find a convex n-gon among them? In other words:

> *What is the maximum number of points in the plane, in general position, that do not contain the vertices of a convex n-gon?*

If we make a table from the known cases, it is not too hard to conjecture the answer:

n	2	3	4	5	6
	1	2	4	8	16?

(We didn't mention the cases for $n = 2$ and 3 before, because these are not interesting, but for the sake of completeness we added them to the table. The "twogon" is nothing but a segment, so it is convex, and every triangle is convex as well.)

We may notice that the number of points not containing a convex n-gon doubles as n increases by 1, at least in the known cases. It is natural to suppose that this will be the truth for larger numbers too, so the conjectured number is 2^{n-2}. It is known that for every $n > 1$, there exists a set of 2^{n-2} points in the plane, in general position, that do not contain a convex n-gon. But whether or not one more point guarantees a convex n-gon is still unknown as of today.

Review Exercises

11.3.2 Given $n \geq 3$ lines in the plane in general position (no two are parallel and no three go through a point), prove that among the regions they divide the plane into, there is at least one triangle.

11.3.3 Into how many parts do two triangles divide the plane?

11.3.4 Into how many parts do two quadrilaterals divide the plane, if

(a) they are convex,

(b) they are not necessarily convex?

11.3.5 Into how many parts do two convex n-gons divide the plane?

11.3.6 Into how many parts can n circles divide the plane, maximum and minimum?

11.3.7 Prove that 6 points in the plane, no 3 on a line, form at least 3 convex quadrilaterals.

11.3.8 Given a set S of 100 points in the plane, no 3 on a line, show that there is a convex polygon whose vertices are in the set S and that contains exactly 50 points of S (including its vertices).

12
Euler's Formula

12.1 A Planet Under Attack

In Chapter 11 we considered problems that can be cast in the language of graph theory: If we draw some special graphs in the plane, into how many parts do these graphs divide the plane? Indeed, we start with a set of lines; we consider the intersections of the given lines as nodes of the graph, and the segments arising on these lines as the edges of our graph. (For the time being, let us forget about the infinite half-lines. We'll come back to the connection between graphs and sets of lines later.)

More generally, we study a *planar map*: a graph that is drawn in the plane so that its edges are nonintersecting continuous curves. We also assume that the graph is connected. Such a graph divides the plane into certain parts, called *countries*. Exactly one country is infinite, the other countries are finite.

A very important result, discovered by Euler, tells us that we may determine the number of countries in a connected planar map if we know the number of nodes and edges of the graph. Euler's Formula is the following:

Theorem 12.1.1 *number of countries + number of nodes = number of edges + 2.*

Proof. To make the proof of this theorem more plausible, we'll tell a little story. This does not jeopardize the mathematical correctness of our proof.

Let us consider the given planar map as the map of a water system of a planet with a single very low continent. We consider the edges not as boundaries between countries but as dams, with watchtowers at the nodes. So the enclosed areas are not countries, but basins. The outermost "basin" is the sea, and all the other "basins" are dry (Figure 12.1). One advantage of this formulation is that we can allow a cut-edge in the graph, which we can think of as a kind of dam, or pier; this could not be considered as a boundary of two countries, since on both sides of it we would have the same "country" (in this case, the sea).

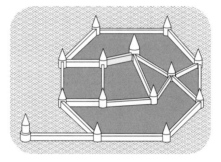

FIGURE 12.1. A graph of dams and watchtowers. There are 14 watchtowers, 7 basins (including the see), and 19 dams. Euler's Formula checks our: $14 + 7 = 19 + 2$.

One day, an enemy attacks the island, and the defenders decide to flood it with seawater by blowing up certain dams. The defenders are hoping to defeat the attack and return to their island, so they try to blow up the smallest possible number of dams. They figured out the following procedure: They blow up one dam at a time, and then only in the case if one side of the dam is already flooded, and the other side is dry. After the destruction of this dam the ocean fills up the previously dry basin with seawater. Notice that all the other dams (edges) around this basin are intact at this stage (because whenever a dam is blown up, the basins on both sides of it are flooded), so the seawater fills up only this particular basin. In Figure 12.2 we have indicated by numbers one possible order in which the dams can be blown up to flood the whole island.

Let us count the number of destroyed and intact dams. We denote the number of watchtowers (nodes) by v, the number of dams (edges) by e, and the number of basins, including the ocean, by f (we'll give an explanation later why are we using these letters). To flood all the $f - 1$ basins of the island, we had to destroy exactly $f - 1$ dams.

To count the surviving dams, let us look at the graph remaining after the explosions (Figure 12.3). First, one can notice that it contains no cycles, because the interior of any cycle would have remained dry. A second observation is that the remaining system of dams forms a connected graph, since every dam that was blown up was an edge of a cycle (the boundary

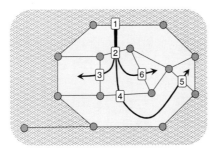

FIGURE 12.2. Flooding the island. To flood 6 basins, 6 dams must be blown up.

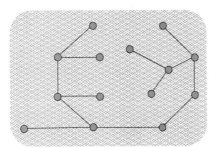

FIGURE 12.3. The island flooded. 13 dams remain intact, forming a tree.

of the basin that was flooded by this last explosion), and we know from exercise 7.2.5(b) that omitting such an edge would not destroy the connectivity of our graph. So the resulting graph after the explosions is connected and does not contain any cycle; therefore, it is a tree.

Now we apply the important fact that if a tree has v nodes, than it has $v - 1$ edges.

Summarizing what we have learned, we know that $f - 1$ dams were blown up and $v - 1$ dams survived. So the number of edges is the sum of these two numbers. Expressing this fact as an equation yields $(v - 1) + (f - 1) = e$. Rearranging, we get

$$f + v = e + 2,$$

and this is just Euler's Formula. □

12.1.1 Into how many parts do the diagonals divide a convex n-gon (we assume that no 3 diagonals go through the same point)?

12.1.2 On a circular island we build n straight dams going from sea to sea, so that every two intersect but no three go through the same point. Use Euler's Formula to determine how many parts we get. As a hint look at Figure 12.4 (the solution of this exercise will be the third method for solving the same problem).

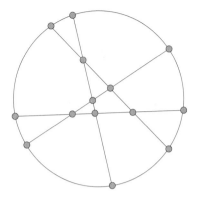

FIGURE 12.4. The planar map given by n straight lines.

12.2   Planar Graphs

Which graphs can be drawn as planar maps? This question is important not only because we want to know to which graphs we can apply Euler's Formula, but also in many applications of graph theory, for example, placing a network on a printed circuit board.

A graph is called *planar* if it can be drawn as a map in the plane, that is, if we can represent its nodes by different points in the plane, and its edges by curves connecting the appropriate points in such a way that these curves don't intersect each other (except, of course, when two edges have a common endpoint, in which case the two corresponding curves will have this one common point).

Are there graphs that are not planar? As a nice application of Euler's Formula, we can prove the following:

Theorem 12.2.1 *The complete graph K_5 on five nodes is not a planar graph.*

One could prove this by distinguishing a large number of cases and using various intuitive but potentially misleading properties of curves in the plane. But we are able to give an elegant proof now, using Euler's Formula.

Proof. Our proof is indirect: Assuming that K_5 can be drawn in the plane without any edges crossing, we get a contradiction. (It is not surprising that we are not able to provide a figure, since the impossibility of such a figure is what we want to prove.) Let us compute the number of countries that we would have in such a drawing. We have 5 nodes and $\binom{5}{2} = 10$ edges; hence the number of countries is, by Euler's Formula, $10 + 2 - 5 = 7$. Every country has at least 3 edges on its boundary, so we must have at least $\frac{3 \cdot 7}{2} = 10.5$ edges. (We had to divide by 2, because we counted every edge for two different countries.) The assumption that K_5 is planar led us to a contradiction, namely, $10 > 10.5$, so our assumption must have been false, and the complete graph on 5 nodes (K_5) is not planar.   □

One of the most interesting phenomena in mathematics occurs when in the proof of some result one can make use of theorems that at a first glance do not have any connection with the actual problem. Would any of you guess that one can use the nonplanarity of the complete graph on five points to give another proof of the starting exercise of Section 11.3? Given our five points in the plane, connect any two of them by a segment. The resulting graph is a complete graph on five vertices, which is not a planar graph, as we already know; therefore, we can find two segments that intersect each other. The four endpoints of these two segments form a quadrilateral whose diagonals intersect; therefore, this quadrilateral is convex.

As another application of Euler's Formula, let's answer the following question: How many edges can a planar map with n nodes have?

Theorem 12.2.2 *A planar graph on n nodes has at most $3n - 6$ edges.*

Proof.   The derivation of this bound is quite similar to our argument above showing that K_5 is not a planar graph. Let the graph have n nodes, e edges, and f faces. We know by Euler's Formula that

$$n + f = e + 2.$$

We obtain another relation among these numbers if we count edges face by face. Each face has at least 3 edges on its boundary, so we count at least $3f$ edges. Every edge is counted twice (it is on the border of two faces), so the number of edges is at least $3f/2$. In other words, $f \leq \frac{2}{3}e$. Using this with Euler's Formula, we get

$$e + 2 = n + f \leq n + \frac{2}{3}e,$$

which after rearrangement gives $e \leq 3n - 6$.   □

12.2.1 Is the graph obtained by omitting an edge of K_5 planar?

12.2.2 There are three houses and three wells. Can we build a path from every house to every well so that these paths do not cross? (The paths are not necessarily straight lines.)

12.3   Euler's Formula for Polyhedra

There is still an apparently irrelevant question to deal with. Why did we denote the number of countries by f? Well, this is the starting letter of the word *face*. When Euler was trying to find "his" formula, he was studying polyhedra (solids bounded by plane polygons) like the cube, pyramids, and prisms. Let us count for some polyhedra the number of faces, edges, and vertices (Table 12.1).

Polyhedron	# of vertices	# of edges	# of faces
cube	8	12	6
tetrahedron	4	6	4
triangular prism	6	9	5
pentagonal prism	10	15	7
pentagonal pyramid	6	10	6
dodecahedron	20	30	12
icosahedron	12	30	20

TABLE 12.1.

(You don't know what the dodecahedron and icosahedron are? These are two very pretty regular polyhedra; their faces are regular pentagons and triangles, respectively. They can be seen in Figure 12.5.)

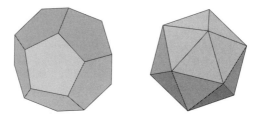

FIGURE 12.5. Two regular polyhedra: the dodecahedron and the icosahedron.

Staring at these numbers for a little while, one discovers that in every case the following relation holds:

number of faces + number of vertices = number of edges + 2.

This formula strongly resembles Euler's Formula; the only difference is that instead of nodes, we speak of vertices, and instead of countries, here we speak of faces. This similarity is not a coincidence; we may get the formula for polyhedra from the formula for planar maps very easily as follows. Imagine that our polyhedron is made out of rubber. Punch a hole into one of the faces and blow it up like balloon. The most familiar solids will be blown up to spheres (for instance the cube and prism). But we have to be careful here: There are solids that cannot be blown up to a sphere. For instance, the "picture frame" shown on Figure 12.6 blows up to a "torus," similar to a life buoy (or doughnut). Be careful, the above relation holds only for solids that can be blown up to spheres! (Just to reassure you, all the convex solids can be blown up to spheres.) Now grab the rubber sphere at the side of the hole and stretch it until you get a huge rubber plane. If we paint the edges of the original solid with black ink, then we will see a map on the plane. The nodes of this map are the vertices of the solid, the edges are the edges of the solid, and the countries are the faces of the body. Therefore, if we use Euler's Formula for maps, we get Euler's Formula for polyhedra (Euler himself stated the theorem in the polyhedral form).

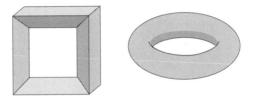

FIGURE 12.6. A nonconvex polyhedron that blows up to a torus.

Review Exercises

12.3.1 Is the complement of the cycle of length 6 (C_6) a planar graph?

12.3.2 Take a hexagon and add the three longest diagonals. Is the graph obtained this way planar?

12.3.3 Does the "picture frame" polyhedron in Figure 12.6 satisfy Euler's Formula?

12.3.4 Prove that a planar bipartite graph on n nodes has at most $2n - 4$ edges.

12.3.5 Using Euler's Formula, show that the Petersen graph is not planar.

12.3.6 A convex polyhedron has only pentagonal and hexagonal faces. Prove that it has exactly 12 pentagonal faces.

12.3.7 Every face of a convex polyhedron has at least 5 vertices, and every vertex has degree 3. Prove that if the number of vertices is n, then the number of edges is at most $5(n - 2)/3$.

12.3.8 Does the "picture frame" polyhedron in Figure 12.6 satisfy Euler's Formula?

13
Coloring Maps and Graphs

13.1 Coloring Regions with Two Colors

We draw some circles on the plane (say, n in number). These divide the plane into a number of regions. Figure 13.1 shows such a set of circles, and also an "alternating" coloring of the regions with two colors; it gives a nice pattern. Now our question is, can we *always* color these regions this way? We'll show that the answer is yes. Let us state this more exactly:

Theorem 13.1.1 *The regions formed by n circles in the plane can be colored with red and blue in such a way that any two regions that share a common boundary arc will be colored differently.*

(If two regions have only one or two boundary points in common, then they may get the same color.)

Let us first see why a direct approach to proving this theorem does not work. We could start with coloring the outer region, say blue; then we have to color its neighbors red. Could it happen that two neighbors are at the same time neighbors of each other? Perhaps drawing some pictures and then arguing carefully about them, you can design a proof that this cannot happen. But then we have to color the neighbors of the neighbors blue again, and we would have to prove that no two of these are neighbors of each other. This could get quite complicated! And then we would have to repeat this for the neighbors of the neighbors of the neighbors. . . .

We should find a better way to prove this, and fortunately, there is a better way!

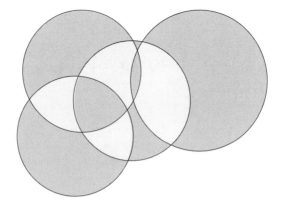

FIGURE 13.1. Two-coloring the regions formed by a set of circles.

Proof. We prove the assertion by induction on n, the number of circles. If $n = 1$, then we get only two regions, and we can color one of them red, the other one blue. So let $n > 1$. Select any of the circles, say C, and forget about it for the time being. We assume that the regions formed by the remaining $n - 1$ circles can be colored with red and blue so that regions that share a common boundary arc get different colors (this is just the induction hypothesis).

Now we put back the remaining circle, and change the coloring as follows: Outside C, we leave the coloring as it was; inside C, we interchange red and blue (Figure 13.2).

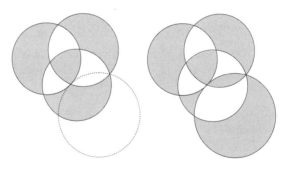

FIGURE 13.2. Adding a new circle and recoloring.

It is easy to see that the coloring obtained this way satisfies what we wanted. In fact, look at any small piece of arc of any of the circles. If this arc is outside C, then the two regions on its two sides were different and their colors did not change. If the arc is inside C, then again, the two regions on its two sides were differently colored, and even though their colors were switched, they are still different. Finally, the arc could be on C itself. Then the two regions on both sides of the arc were one and the same before we put C back, and so they had the same color. Now, one of them is inside C and this switched its color; the other is outside, and this did not. So after the recoloring, their colors will be different.

Thus we have proved that the regions formed by n circles can be colored with two colors, provided that the regions formed by $n-1$ circles can be colored with 2 colors. By the Principle of Induction, this proves the theorem.   □

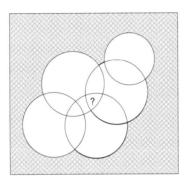

FIGURE 13.3. Find the color of a region.

13.1.1 Assume that the color of the outer region is blue. Then we can describe what the color of a particular region R is without having to color the whole picture as follows:

— if R lies inside an even number of circles, it will be colored blue;

— if R lies inside an odd number of circles, it will be colored red.

Prove this assertion (see Figure 13.3).

13.1.2 (a) Prove that the regions, into which n straight lines divide the plane, are colorable with 2 colors.

(b) How could you describe what the color of a given region is?

13.2 Coloring Graphs with Two Colors

Jim has six children, and it is not an easy bunch. Chris fights with Bob, Faye, and Eve all the time; Eve fights (besides with Chris) with Al and Di all the time; and Al and Bob fight all the time. Jim wants to put the children in two rooms so that pairs of fighters should be in different rooms. Can he do this?

As in Chapter 7, the solution is much easier if we make a drawing of the information we have. We construct a "fighting graph": We represent each child by a node, and connect two nodes if these children fight all the time. We get the graph in Figure 13.4(a). Jim's task is to split the nodes into two groups so that every fighting pair is separated. A solution is shown in part (b) of the figure.

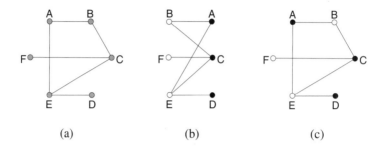

(a)   (b)   (c)

FIGURE 13.4. The fight graph of Jim's children; how to put them in two rooms; and how this translates to a 2-coloring.

Instead of redrawing the graph putting its nodes left and right, we can indicate which child goes in which room by coloring the corresponding node black or white. The rule of the coloring is that *adjacent nodes must be colored with different colors*. Figure 13.4(c) shows the corresponding coloring. Jim can put the children represented by white nodes in one room, and the others (represented by black nodes) in the other.

Looking at Figure 13.4(b), we notice that we have already met such graphs: we called them bipartite, since the vertex set of these graphs can be split into two disjoint sets (or parts) such that edges go only between vertices belonging to different sets.

The problem of coloring the regions formed by circles in the previous section can also be stated as a problem on coloring the nodes of graphs with 2 colors. We associate a graph with the regions formed by circles the following way: Represent every region by a vertex of the graph. Two vertices are connected by an edge in the graph if and only if the corresponding regions share a common boundary arc (just sharing a point at a vertex does not qualify).

Which graphs are 2-colorable (in other words, bipartite)? It is clear that if a graph consists of isolated vertices, 1 color is enough to get a good coloring. If the graph has at least one edge, then we need at least two colors. It is easy to see that a triangle, the complete graph on 3 vertices, needs 3 colors to be well colored. It is also obvious that if a graph contains a triangle, then it needs at least 3 colors for a good coloring.

But a graph need not contain a triangle and may still not be 2-colorable. A little more involved example is a pentagon: It is easy to convince ourselves that no matter how we color its vertices with 2 colors, we'll necessarily end up with two adjacent vertices with the same color. We can generalize this observation to any cycle of odd length: If we start coloring any node with (say) black, then as we walk along the cycle we must color the next node white, the third node black, and so on. We must alternate with colors black and white. But because the cycle is odd, we return to the start node in the wrong phase, ending up with coloring the last node black, and so having two adjacent black nodes.

It follows that if a graph contains an odd cycle, then it cannot be 2-colorable. The following simple theorem asserts that nothing more complicated can go wrong:

Theorem 13.2.1 *A graph is 2-colorable if and only if it contains no odd cycle.*

Proof. We already know the "only if" part of this theorem. To prove the "if" part, suppose that our graph has no odd cycle. Pick any vertex a and color it black. Color all its neighbors white. Notice that there cannot be an edge connecting two neighbors of a, because this would give a triangle. Now color every uncolored neighbor of these white vertices black. We have to show that there is no edge between the black vertices: no edge goes between u and the new black vertices, since the new black vertices didn't belong to the neighbors of a; no edge can go between the new black vertices, because it would give a cycle of length 3 or 5. Continuing this procedure the same way, if our graph is connected, we'll end up with 2-coloring all vertices.

It is easy to argue that there is no edge between two vertices of the same color: Suppose that this is not the case, so we have two adjacent vertices u and v colored black (say). The node u is adjacent to a node u_1 colored earlier (which is white); this in turn is adjacent to a node u_2 colored even earlier (which is black); etc. This way we can pick a path P from u that goes back all the way to the starting node. Similarly, we can pick a path Q from v to the starting node. Starting from v, let's follow Q back until it first hits P, and then follow P forward to u. This path forms a cycle with the edge uv. Since the nodes along the path alternate in color, but start and end with black, this cycle is odd, a contradiction (Figure 13.5).

If the graph is connected, we are done: We have colored all vertices. If our graph is not connected, we perform the same procedure in every component, and obviously, this will give a good 2-coloring of the whole graph. This proves Theorem 13.2.1.   □

It is worth pointing out that in the proof above we did not really have much choice: We could choose the color of the starting node, but then our hands were forced all the way through coloring the connected component of that point. Then we had a free choice for the color of the first node of the next component, but then the rest of the component was forced again. This means that we not only proved Theorem 13.2.1, but also gave an *algorithm* for finding a 2-coloring if it exists (and to find an odd cycle if it does not).

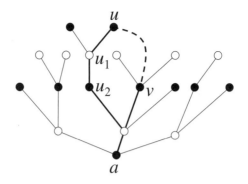

FIGURE 13.5. Bad 2-coloring yields an odd cycle.

13.2.1 Prove that the graph obtained from the regions of a system of circles contains no odd cycle.

13.3 Coloring graphs with many colors

Suppose that we have a graph, and we find that it cannot be colored with two colors; we may want to color it with more colors. (The rule of the game is the same: the two endpoints of any edge must be colored differently.) If we have many colors, then we can just color every node with a different color. If the graph has n nodes, then n colors are always sufficient. Obviously, if the graph is complete, then we do need n colors, since every node must get a different color (recall the Pigeonhole Principle!).

If we can color a graph with k colors, we say that the graph is k-*colorable*. The smallest k for which the graph is k-colorable is called the *chromatic number* of the graph.

Suppose that we want to use, say, only 3 colors. Can we decide whether they are enough to color the nodes? It turns out that going from 2 colors to 3 is a real jump in difficulty.

We can try to proceed as in the case of 2 colors. Let the 3 colors be red, blue and green. We start with any node a, and color it red. This we can do without loss of generality, since all colors play the same role. Then we take a neighbor b of a, and color it blue (this is still no restriction of generality). Now we proceed to another neighbor c of a. If c is connected to b by an edge, then its color is forced to be green. But suppose that it is not connected to b. Then we have two choices to color c, blue or green, and these are *not* alike: It makes a difference whether b and c have the same color or not. So we make a choice and proceed. In the next step, we may have to make a choice again, etc. If any of these choices turns out to be wrong, we'll have to backtrack and try the other one. Eventually, if a 3-coloring is possible at all, we'll find it.

All this backtracking takes a lot of time. We don't give a rigorous analysis here, just note that in general, we may have to go through both choices for a large fraction (say, half) of the nodes, which will take time $2^{n/2}$. We have seen that this number becomes astronomical for very moderate sizes of n.

But the jump in difficulty is not that this simple procedure takes so long: the real trouble is that nothing substantially better is known! And there are results in complexity theory (cf. Section 15.1) that suggest that nothing substantially better can be designed at all. The situation is similar for coloring with any number k of colors, once $k > 2$.

Suppose that we have a graph and we badly need to color it with k colors. Are there at least some special cases when we can do so? One such case is described in the following result, called Brooks's Theorem:

Theorem 13.3.1 *If every node in a graph has degree at most d, then the graph can be colored with $d + 1$ colors.*

Of course, the condition given in Brooks's Theorem is only sufficient, but not necessary: there are graphs in which some nodes, possibly even all nodes have high degree, and the graph is colorable with 2 colors.

Proof. We give a proof using the Principle of Induction. We can start our proof with small graphs. If the graph has less than $d + 1$ vertices, then it can be colored by $d + 1$ (or fewer) colors, since every vertex can be colored differently. Suppose that our theorem is true for any graph with fewer than n vertices. Pick a point v of our graph G, and omit from G this vertex v and the edges incident to it. The remaining graph G' has $n - 1$ vertices. Obviously every degree is at most d (omitting a vertex does not increase the degrees), so G' can be colored by $d + 1$ colors, by the induction hypothesis. Since v has at most d neighbors, but we have $d + 1$ colors, there must be a color that does not occur among the neighbors of v. Coloring v with this color, we get a coloring of G by $d + 1$ colors. This completes the induction. □

Brooks, in fact, proved more: With some simple exceptions, a graph in which all nodes have degree at most d can be colored with d colors. The exceptions can be described as follows. For $d = 2$, the graph has a connected component that is an odd cycle; for $d > 2$, the graph has a connected component that is a complete graph with $d + 1$ nodes. The proof that these graphs are exceptions is easy; the proof that these are the only exceptions is harder and not given in this book.

Returning to the situation where you want to k-color a graph, suppose you find that Brooks's Theorem does not apply, and say a large number of random trials to color the graph also fail, and you begin to suspect that no k-coloring exists at all. How can you convince yourself that this is indeed the case? One lucky break is if you find $k + 1$ nodes in the graph forming a complete subgraph. Obviously, this part of the graph needs $k + 1$ colors.

Unfortunately, for every positive integer k there are graphs that contain no such complete graph and yet they are not k-colorable. It may even happen that a graph contains no triangle and it is not k-colorable. Figure 13.6 shows a graph that contains no complete graph with 4 vertices but is not 3-colorable, and a more complicated graph that does not contain a triangle but is not 3-colorable (see also Exercise 13.3.1).

On this sad note we leave the topic of coloring general graphs.

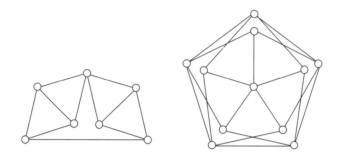

FIGURE 13.6. Two non-3-colorable graphs.

13.3.1 Prove that the graphs in Figure 13.6 are not 3-colorable.

13.3.2 Draw n lines in the plane such that no 3 of them go through the same point. Prove that their intersection points can be colored with 3 colors so that on every line, consecutive intersection points get different colors.

13.3.3 Let G be a connected graph such that all vertices but one have degree at most d (one vertex may have degree larger than d). Prove that G is $(d+1)$-colorable.

13.3.4 Prove that if every subgraph of G has a node of degree at most d, then G is $(d+1)$-colorable.

13.4 Map Coloring and the Four Color Theorem

We started this chapter by coloring the regions formed by a set of circles in the plane. But when do we need to color drawings in the plane? Such a task arises in cartography: It is a natural requirement to color maps in such a way that neighboring countries (or counties) get different colors. In the previous example we saw that in special cases (like maps derived from circles), we may find "good" colorings of the maps in the previous sense, using just two colors. "Real" maps are much more complicated configurations, so it is not surprising that they need more than two colors. It is very easy to draw four countries so that any two of them have a common

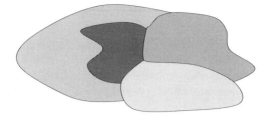

FIGURE 13.7. Four mutually neighboring countries

boundary, and so all four need different colors in a "good" coloring (see Figure 13.7).

Now consider a "real-life" planar map, for instance the map of the states of the continental US. We assume that each country (state) is connected (consists of one piece). In school maps usually six colors are used, but four colors are enough, as shown in Figure 13.8.

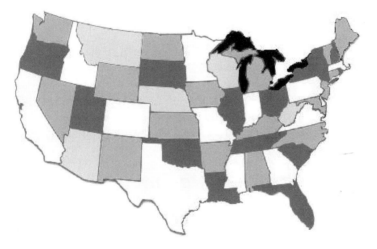

FIGURE 13.8. Coloring the states of the United States.

Would three colors be enough? One can easily see that the answer is negative. Let's start to color our map with three colors, say red, blue, and yellow (it does not make any difference which of the three colors we start with); we have to prove that sooner or later we get stuck. Let's start by coloring Nevada red. Then all the neighboring states can get only colors blue or yellow. Let's color California blue (without loss of generality). Then the next neighboring state clockwise, Oregon, is a common neighbor of both Nevada and California, so it must be colored yellow; the next state, Idaho, must be blue again; the next state, Utah, must be yellow. But now we are stuck, because the last state among the neighbors of Nevada, Arizona, cannot get any of the three colors red, blue, or yellow, so we must use a fourth color (see Figure 13.8).

It is not by accident that in two different cases to color a map we needed four colors, but four colors were enough. It is a theorem that *to color any planar map, four colors always suffice.*

This famous theorem has a history well over a century old. It was raised by Francis Guthrie in England in 1852. For decades it was kicked around by mathematicians as a simple but elusive puzzle, until the difficulties in obtaining a proof became apparent in the 1870s. An erroneous proof was published in 1879 by Alfred Kempe , and the problem was regarded as solved for a good decade before the error was discovered. The difficulty of the problem was overlooked to the degree that in 1886 it was posed at Clifton College as a challenge problem to students; part of the requirement was that "No solution may exceed one page, 30 lines of MS., and one page of diagrams." (The real solution 90 years later used more than 1000 hours of CPU time!)

After the collapse of Kempe's proof, for more than a century many mathematicians, amateur and professional, tried in vain to solve this intriguing question, called the *Four Color Conjecture.* Several further erroneous proofs were published and the refuted.

A whole new area of mathematics, graph theory, grew out of attempts to prove the Four Color Conjecture. Finally, in 1976 events took a surprising turn: Kenneth Appel and Wolfgang Haken gave a proof of the Four Color Conjecture (which is therefore the Four Color Theorem now), but their proof used computers very heavily to check an enormous number of cases. Even today, the use of computers could not be eliminated from the proof (although nowadays it takes much less time than the first proof, partly because computers are faster, partly because a better arrangement of the case distinction was found); we still don't have a "pure" mathematical proof of this theorem.

It is beyond the scope of this book even to sketch this proof; but we can use the results about graphs that we have learned to prove the weaker fact that *5 colors suffice to color every planar map.*

We can transform this to a graph coloring problem. In each country we pick a point; let's call this the *capital* of the country. Now, if two countries share a border, then we can connect their capitals by a railway line that stays in the two countries and crosses the border only once. Furthermore, we can design these lines so that the lines from any capital to various points on the border of the country (going on to capitals of adjacent countries) do not cross each other. Then the capitals and the lines connecting them form a graph, which also is planar. This graph is called the *dual graph* of the original map (Figure 13.9).

A word here to make things precise: It could happen that the common border of two countries consists of several pieces. (For example, in our world the border of China and Russia consists of two pieces, separated by

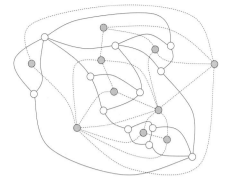

FIGURE 13.9. A graph and its dual

Mongolia.) For the purpose of studying colorings, it suffices to connect the capitals of these two countries by one line only, through either one of these pieces of their common border.

There is another graph in the picture, consisting of the borders between countries. The nodes of this graph are the "triangles," points where three or more countries meet. But we have to be careful: This "graph" may have two or more edges connecting the same pair of nodes! So this is an example where we need multigraphs to model the situation correctly. We don't need to bother about this, however: we can just talk about a planar map and its dual graph.

Instead of coloring the countries of the original map, we could color the nodes of its dual graph: Then the rules of the game would be that two nodes connected by an edge must be colored with different colors. In other words, this is graph coloring as defined in Section 13.3. So we can rephrase the Four Color Theorem as follows:

Theorem 13.4.1 *Every planar graph can be colored with 4 colors.*

Our more modest goal is to prove the "Five Color Theorem" :

Theorem 13.4.2 *Every planar graph can be colored with 5 colors.*

Let us start with something even more modest:

Every planar graph can be colored with 6 colors.

(7 colors or 8 colors would be even more modest, but would not be any easier.)

Let us look at what we already know about graph coloring; is any of it applicable here? Do we know any condition that guarantees that the graph is 6-colorable? One such condition is that all points in a graph have degree at most 5 (Theorem 13.3.1). This result is not applicable here, though, because a planar graph can have points of degree higher than 5 (the "dual" graph in Figure 13.9 has a point of degree 7, for example). But if you solved the exercises, you may recall that we don't have to assume that *all* nodes of the graph have degree at most 5. The same procedure as used in the proof of Theorem 13.3.1 gives a 6-coloring if we know that the graph has *at least one* point of degree 5 or less, and so do all its subgraphs (Exercise 13.3.4). Is this condition applicable here?

The answer is yes:

Lemma 13.4.3 *Every planar graph has a point of degree at most 5.*

Proof. This lemma follows from Euler's Formula. In fact, we only need a consequence of Euler's Formula, namely, Theorem 12.2.2: *A planar graph with n nodes has at most $3n - 6$ edges.*

Assume that our graph violates Lemma 13.4.3, and so every node has degree at least 6. Then counting the edges node by node, we count at least $6n$ edges. Each edge is counted twice, so the number of edges is at least $3n$, contradicting Theorem 12.2.2.   □

Since the subgraphs of a planar graph are planar as well, it follows that they too have a point of degree at most 5, and so Exercise 13.3.4 can be applied, and we get that every planar graph is 6-colorable.

So we have proved the "Six Color Theorem." We want to shave off 1 color from this result (how nice it would be to shave off 2!). For this, we use the same procedure of coloring points one by one again, together with Lemma 13.4.3; but we have to look at the procedure more carefully.

Proof [of the Five Color Theorem]. So, we have a planar graph with n nodes. We use induction on the number of nodes, so we assume that planar graphs with fewer than n nodes are 5-colorable. We also know that our graph has a node v with degree at most 5.

If v has degree 4 or less, then the argument is easy: Let us delete v from the graph, and color the remaining graph with 5 colors (which is possible by the induction hypothesis, since this is a planar graph with fewer nodes). The node v has at most 4 neighbors, so we can find a color for v that is different from the colors of its neighbors, and extend the coloring to v.

So the only difficult case occurs when the degree of v is exactly 5. Let u and w be two neighbors of v. Instead of just deleting v, we change the graph a bit more: We use the place freed up by the deletion of v to merge u and w to a single point, which we call uw. Every edge that entered either u or w will be redirected to the new node uw (Figure 13.10).

This modified graph is planar and has fewer nodes, so it can be colored with 5 colors by the induction hypothesis. If we pull the two points u and w apart again, we get a coloring of all nodes of G except v with 5 colors. What we gained by this trick of merging u and w is that in this coloring they have the same color! So even though v has 5 neighbors, two of those have

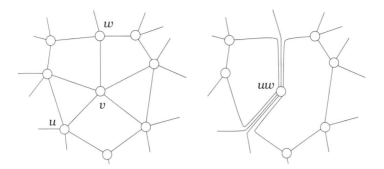

FIGURE 13.10. Proof of the 5-color theorem.

the same color, so one of the 5 colors does not occur among the neighbors at all. We can use this color as the color of v, completing the proof.

Warning! We have overlooked a difficulty here. (You can see how easy it is to make errors in these kinds of arguments!) When we merged u and w, two bad things could have happened: (a) u and w were connected by an edge, which after the merging became an edge connecting a node to itself; (b) there could have been a third node p connected to both u and w, which after the merging became a node connected to uw by two edges. We did not allow for either of these!

The second trouble (b) is in fact no trouble at all. If we get two edges connecting the same pair of nodes, we could just ignore one of them. The graph remains planar, and in the 5-coloring the color of p would be different from the common color of u and w, so when we pull them apart, both edges connecting p to u and w would connect nodes with different color.

But the first trouble (a) is serious. We cannot just ignore the edge connecting uw to itself; in fact, there is no way that u and w can get the same color in the final coloring, since they are connected by an edge!

What comes to the rescue is the fact that we can choose another pair u, w of neighbors of v. Could it happen that we have this problem with every pair? No, because then every pair of neighbors would be adjacent, and this would mean a complete graph with 5 nodes, which we know is not planar. So we can find at least one pair u and w for which the procedure above works. This completes the proof of the Five Color Theorem.   □

Review Exercises

13.4.1 We draw a closed curve in the plane without lifting the pencil, intersecting itself several times (Figure 13.11). Prove the fact (familiar from boring classes) that the regions formed by this curve can be colored with 2 colors.

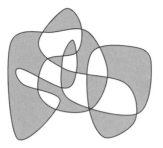

FIGURE 13.11. 2-coloring the regions of a curve

13.4.2 Let G be a connected graph such that all vertices have degree at most d, and there exists vertex with degree strictly less than d. Prove that G is d-colorable.

13.4.3 Let G be a connected graph such that all vertices but $d+1$ have degree at most d (the remaining $d+1$ vertices may have degree larger than d). Prove that G is $(d+1)$-colorable.

13.4.4 Construct a graph G as follows: The vertices of G are the edges of a complete graph K_5 on 5 vertices. The vertices of G are adjacent if and only if the corresponding edges of K_5 have an endpoint in common. Determine the chromatic number of this graph.

13.4.5 Let G_n be the graph arising from K_n (where K_n is the complete graph on n vertices) by omitting the edges of a Hamiltonian cycle. Determine the chromatic number of G_n.

13.4.6 Show by an example that on a continent where countries are not necessarily connected (as in medieval Europe), 100 colors may not be enough to color a map.

13.4.7 If every face of a planar map has an even number of edges, then the graph is bipartite.

13.4.8 If every node of a planar map has even degree, then the faces can be 2-colored.

13.4.9 (a) Consider a planar map in which every node has degree 3. Suppose that the faces can be 3-colored. Prove that the graph of the map is bipartite.
   (b) [A challenge exercise.] Prove the converse: If every node of a bipartite planar graph has degree 3, then in the map obtained by drawing it in the plane, the faces can be 3-colored.

14

Finite Geometries, Codes, Latin Squares, and Other Pretty Creatures

14.1 Small Exotic Worlds

The Fano plane is a really small world (Figure 14.1). It only has 7 points, which form 7 lines. In the figure, 6 of these lines are straight and one is circular; but for the inhabitants of this tiny world (the Fanoans), the straight and curved lines look the same. Also, in our figure it seems that the circular line intersects some of the straight lines twice; but the intersection points that are not marked are not real intersection points, they are there only because we want to draw a picture of this very different world in our Euclidean plane.

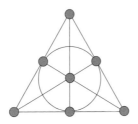

FIGURE 14.1. The Fano plane.

The Fanoans are very proud of their world. They say that it is tiny but perfect in various ways. They point out that

(a) *through any two points their world has a unique line.*

If you check this out and admit it's true, but reply that this also holds in our own world, they go on and boast that

(b) *any two lines have exactly one intersection point.*

This is certainly not true in our Euclidean world (we have parallel lines), so we have to admit that this is nice indeed. But then we can draw a new figure (Figure 14.2) and point out that this rather uninteresting construction also has properties (a) and (b). But the Fanoans are ready for this attack: "It would be enough if we pointed out that in our world,

(c) *every line has at least 3 points.*"

FIGURE 14.2. An ugly plane

Our Fanoan friend goes on: "Theoretical physicists have shown that just from (a), (b), and (c), many properties of our world can be derived. For example,

(d) *all our lines have the same number of points.*

"Indeed, let K and L be two lines. By (b), they have an intersection point p; by (c), they contain other points (at least two, but we only need one right now). Let us select a point q from K and a point r from L that are both different from p. By (a), there is a line M through q and r. Let s be a third point on M (which exists by (c); see Figure 14.3).

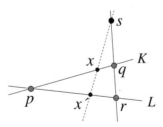

FIGURE 14.3. All lines have the same number of points.

"Consider any point x on K, and connect it to s. This line intersects L at a point x' (we can think of this as projecting K onto L from center s). Conversely, given x', we get x by the same construction. Thus this projection establishes a bijection between K and L, and so they have the same number of points.

"This argument also shows that

(e) *all our points have the same number of lines through them.*

"No doubt you can prove this yourself, if you study the previous argument carefully."

Our intelligent Fanoan friend adds, "These theoretical physicists (obviously having a lot of time on their hands) also determined that the fact that the number of points on each line is 3 does *not* follow from (a), (b), and (c); they say that alternative universes can exist with 4, 5 or 6 points on a line. Imagining this is beyond me, though! But they say that no universe could have 7 points on a line; there could be universes with 8, 9, and 10 points on each line, but 11 points are impossible again. This is, of course, a favorite topic for our science fiction writers."

The Fanoans hate Figure 14.2 for another reason: they are true egalitarians, and the fact that one point is special is intolerable in their society. You may raise here that the Fano plane also has a special point, the one in the middle. But they immediately explain that this is again an artifact of our drawing. "In our world,

(f) *all points and all lines are alike*

in the sense that if we pick any two points (or two lines), we can just rename every point so that one of them becomes the other, and nobody will notice the difference." You may trust them about this, or you may verify this claim by solving exercise 14.1.7.

Let us leave the Fano plane now and visit a larger world, the *Tictactoe plane*. This has 9 points and 12 lines (Figure 14.4). We have learned from our excursion to the Fano plane that we have to be careful with drawing these strange worlds, and so we have drawn it in two ways: In the second figure, the first two columns are repeated, so that the two families of lines (one leaning right, one leaning left) can be seen better.[1]

The Tictacs boast that they have a much more interesting world than the Fanoans. It is still true that any two points determine a single line; but two lines may be intersecting or parallel (which simply means that they don't intersect). One of our Tictac friends explains: "I heard that your mathematicians have been long concerned about the statement that

(g) *for any line and any point not on the line, there is one and only one line that goes through the point and is parallel to the given line.*

They called it the Axiom of Parallels or Euclid's Fifth Postulate. They were trying to prove it from other basic properties of your world, until eventually

---

[1]If you have learned about matrices and determinants, you may recognize the following description of this world: if we think of the points in this plane as the entries of a 3×3 matrix, then the lines are the rows and columns of the matrix and expansion terms of its determinant. The second drawing in the figure corresponds to Sarrus's Rule in the theory of determinants.

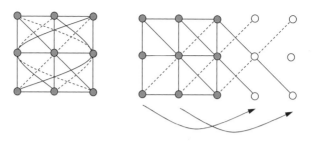

FIGURE 14.4. The Tictactoe plane.

they showed that this cannot be done. Well, this is true in our world, and since our world is finite, it is easy to check that it is true." (We hope our readers will accept the challenge.)

"We have 3 points on each line just as in the Fano plane, but we have 4 lines through each point—more than those Fanoans. All of our points are alike, and so are all of our lines (even though the way you draw them in your own geometry seems to differentiate between straight and curved lines)."

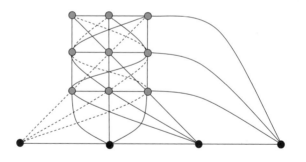

FIGURE 14.5. Extending the Tictactoe plane by 1 line and 4 points at infinity.

When we point out how the Fanoans love their property (b), our Tictac guide replies, "We could easily achieve this ourselves. All we'd have to do is add 4 new points to our plane. Each line should go through exactly one of these new points; parallel lines should go through the same new point, nonparallel lines through different new points. And we could even things out by declaring that the 4 new points also form a line. We could call the new points 'points at infinity' and the new line, the 'line at infinity' (Figure

14.5). Then we would have properties (a) through (g) ourselves.[2] But we prefer to distinguish between finite and infinite points, which makes our world more interesting."

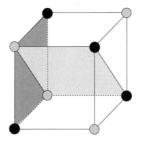

FIGURE 14.6. The Cube space.

Finally, we visit a third tiny world called the *Cube space* (Figure 14.6). While this one has only 8 points, it is much richer than the Tictactoe plane in one sense: It is 3-dimensional! Its lines are uninteresting: Every line has just 2 points, and any two points form a line. But it has planes! In our (deficient) Euclidean picture, the points are arranged as the vertices of a cube. The planes are (1) 4-tuples of points forming a face of the cube (there are six of these), (2) 4-tuples of points on two opposite edges of the cube (6 of these again), (3) the four black points, and (4) the four light points.

The Cube space has the following very nice properties (whose verification is left to the reader):

(A) *Any three points determine a unique plane.*

(The Cubics remark at this point, "In your world, this is only true if the three points are not on a line. Luckily, we never have three points on a line!")

(B) *Any two planes are either parallel (nonintersecting), or their intersection is a line.*

(C) *For any plane and any point outside it, there is exactly one plane through the given point parallel to the given plane.*

(D) *Any two points are alike.*

(E) *Any two planes are alike.*

This last claim looks so unlikely, considering that we have such different kinds of planes, that a proof is in order. Let us label the points of the cube by $A, \ldots, H$ as in Figure 14.7. It is clear that the faces of the cube are alike

---

[2]This construction appending new points at infinity can be carried out in our own Euclidean plane, leading to an interesting kind of geometry, called *projective geometry*.

(one can be moved onto any other by appropriately rotating the cube, and this rotation maps all the other planes onto planes). It is also clear that the planes formed by two opposite edges are alike, and the all-black plane is like the all-light plane (reflecting the cube in its center will interchange black and light vertices).

This was the easy part. Now we do a trickier transformation: We interchange E with F and G with H (Figure 14.7). What happens to the planes?

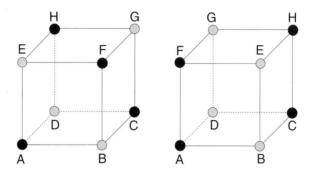

FIGURE 14.7. Why different-looking planes are alike.

Some of them are not changed (even though their points will change places): The top, bottom, front, and back faces and the planes $ABGH$ and $CDEF$ are mapped onto themselves. The left face $ADEH$ is mapped onto the plane $ADFG$, and vice versa. Similarly, the right face $BCFG$ is mapped onto the plane $BCEH$, and vice versa. The all-black plane is mapped onto the plane $ACEG$, and vice versa. The all-light plane is mapped onto the plane $BDFH$, and vice versa.

Thus all planes are accounted for. We make two observations:

— *Every plane is mapped onto a plane*, and so if we relabel the cube as above, the Cubics won't notice the difference!

— There is a face-plane mapped onto an opposite-edge-plane, and there is an opposite-edge-plane mapped onto the all-black plane. This implies that the Cubics cannot see any difference between these three types of planes.

14.1.1 Fanoan philosophers have long been troubled by the difference between points and lines. There are many similarities (for example, there are 7 of each); why are they different? Represent each line by a new point; for each old point, take the 3 lines through it, and connect the 3 new points representing them by a (new) line. What structure do you get?

14.1.2 The Fanoans call a set of 3 points a *circle* if they are not on a line. For example, the 3 vertices in figure 14.1 form a circle. They call a line a *tangent* to the circle, if it contains exactly one point of the circle. Show that at every point of a circle there is exactly one tangent.

14.1.3 The Fanoans call a set of 4 points a *hypercircle* if no 3 of them are a line. Prove that the 3 points not on a hypercircle form a line, and vice versa.

14.1.4 Representatives of the 7 points in the Fano plane often vote on different issues. In votes where everyone has to vote yes or no, they have a strange rule to count ballots, however: it is not the majority who wins, but "line wins": if all 3 points on a line want something, then this is so decided. Show that (a) it cannot happen that contradictory decisions are reached because the points on another line want the opposite, and (b) in every issue there is a line whose points want the same, and so decision is reached.

14.1.5 Prove that the Tictactoe plane, extended with elements at infinity, satisfies all properties (a)–(d).

14.1.6 In response to the Tictacs' explanation about how they could extend their world with infinite elements, the Fanoans decided to declare one of their lines the "line at infinity," and the points on this line "points at infinity." The remaining 4 points and 6 lines form a really tiny plane. Will property (g) of parallel lines be valid in this geometry?

14.1.7 We want to verify the claim of the Fanoans that all their points are alike, and rearrange the points of the Fano plane so that the middle point becomes (say) the top point, but lines remain lines. Describe a way to do this.

14.1.8 Every point of the Cube space is contained in 7 lines and 7 planes. Is this numerical similarity with the Fano plane a coincidence?

14.2   Finite Affine and Projective Planes

It is time to leave our excursion to imaginary worlds and introduce mathematical names for the structures we studied above. If we have a finite set V whose elements are called *points*, and some of its subsets are called *lines*, and (a), (b), and (c) above are satisfied, then we call it a *finite projective plane*. The Fano plane (named after the Italian mathematician Gino Fano) is one projective plane (we'll see that it has the least possible number, 7, of points), and another one was constructed by the Tictac theoretical physicists by adding to their world 4 points and a line at infinity.

We have heard the proof from Fanoan scientists that every line in a finite projective plane has the same number of points; for reasons that should become clear soon, this number is denoted by $n + 1$, where the positive integer n is called the *order* of the plane. So the Fano plane has order 2, and the extended Tictactoe plane has order 3. The Fanoans also know that if a finite projective plane has order n, then $n + 1$ lines go through every point of it.

14.2.1 Prove that a finite projective plane of order n has $n^2 + n + 1$ points and $n^2 + n + 1$ lines.

We can also study structures consisting of points and lines, where (a) and the "Axiom of Parallels" (g) are assumed; to exclude trivial (ugly?) examples, we assume that each line has at least 2 points. Such a structure is called a *finite affine plane*.

The "Axiom of Parallels" implies that all lines parallel to a given line L are also parallel to each other (if two of them had a point p in common, then we would have two lines through p parallel to L). So all lines parallel to L form a "parallel class" of mutually parallel lines, which cover every point in the affine plane.

Affine versus projective planes. The construction used by the Tictacs to extend their plane can be used in general. To every parallel class of lines we append a new "point at infinity" and create a new "line at infinity" going through all points at infinity. Then (a) remains satisfied: Two "ordinary" points are still connected by a line (the same line as before), two "infinite" points are connected by a line (the "infinite" line), and an ordinary and an infinite point are connected by a line (the parallel class belonging to the infinite point contains a line through the given ordinary point). It is even easier to see that we do not have two lines through any pair of points.

Furthermore, (b) is satisfied: Two ordinary lines intersect each other unless they are parallel, in which case they share a point at infinity; an ordinary line intersects the infinite line at its point at infinity. We leave it to you to check (c), (d), and (e) (condition (f) does not hold for every finite projective plane; it is a special feature of the Fano plane and some other projective planes).

The construction in Exercise 14.1.6 is again quite general. We can take any finite projective plane, and call any of its lines along with the points on this line "infinite." The remaining points and lines form a finite affine plane. So in spite of the rivalry between the Fanoans and Tictacs, finite affine planes and projective planes are essentially the same structures.

To sum up, we have the following theorem.

Theorem 14.2.1 *Every finite affine plane can be extended to a finite projective plane by adding new points and a single new line. Conversely, from every projective plane we can construct an affine plane by deleting any line and its points.*

A projective plane of order n has $n + 1$ points on each line; the corresponding affine plane has n. We call this number the *order* of the affine plane. (So this turns out to be more natural for affine planes than for projective planes. We'll see soon why we chose the number of points on a line of the affine plane, rather than on a line of the projective plane, to be called the order.)

We have seen (Exercise 14.2.1) that the projective plane has $n^2 + n + 1$ points. To get the affine plane, we delete the $n + 1$ points on a line, so the affine plane has n^2 points.

Coordinates. We have discussed two finite planes (affine or projective; we know it does not matter much): the Fano and the Tictactoe planes. Are there any others?

Coordinate geometry gives the solution: Just as we can describe the Euclidean plane using real coordinates, we can describe finite affine planes using the strange arithmetic of prime fields from Section 6.8. Let us fix a prime p, and consider the "numbers" (elements of the prime field) $\overline{0}, \overline{1}, \ldots, \overline{p-1}$.

In the Euclidean plane, every point has two coordinates, so let's do the same here: Let the points of the plane be all pairs $(\overline{u}, \overline{v})$. This gives us p^2 points.

We have to define the lines. In the Euclidean plane, these are given by linear equations, so let's do the same here: For every equation

$$\overline{a}x + \overline{b}y = \overline{c},$$

we take the set of all pairs $(\overline{u}, \overline{v})$ for which $x = \overline{u}$, $y = \overline{v}$ satisfies the equation, and introduce a line containing all these points. To be precise, we have to assume that the above equation is proper, in the sense that at least one of $\overline{a}$ and $\overline{b}$ is different from 0.

Now we have to verify that (a) through any two points there is exactly one line, (b) for any line and any point outside it, there is exactly one line through the point that is parallel to the line, and (c) there are at least 2 points on each line. We'll not go through this proof, which is not difficult, but lengthy. It is more important to realize that *all this works because it works in the Euclidean plane, and we can do arithmetic in prime fields just as with real numbers.*

This construction provides an affine plane for every prime order (from this we can construct a projective plane for every prime order). Let's see what we get from the smallest prime field, the 2-element field. This will have $2^2 = 4$ points, given by the four pairs $(0,0), (0,1), (1,0), (1,1)$. The lines will be given by linear equations, of which there are six: $x = 0$, $x = 1$, $y = 0$, $y = 1$, $x + y = 0$, $x + y = 1$. Each of these lines goes through 2 points; for example, $y = 1$ goes through $(0,1)$ and $(1,1)$, and $x + y = 0$ goes through $(0,0)$ and $(1,1)$. So we get the very trivial affine plane (already familiar from Exercise 14.1.6) consisting of 4 points and 6 lines. If we extend this to a projective plane, we get the Fano plane.

Figure 14.8 shows the affine plane of order 5 obtained this way (we don't show all the lines; there are too many).

14.2.2 Show that the Cube space can be obtained by 3-dimensional coordinate geometry from the 2-element field.

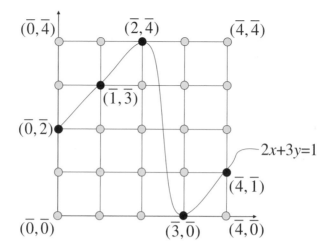

FIGURE 14.8. An affine plane of order 5. Only one line is shown besides the trivial "vertical" and "horizontal" lines.

What are the possible orders of planes? The construction with coordinates above shows that for every prime number there is a finite affine (or projective) plane of that order. Using similar but more involved algebra, one can construct projective planes for every order that is a higher power of a prime number (so for 4, 8, 9 etc.)

Theorem 14.2.2 *For every order that is a power of a prime (including the primes themselves) there is a finite affine (as well as a finite projective) plane of that order.*

The smallest positive integer that is not a prime power is 6, and Gaston Tarry proved in 1901 that no finite plane of order 6 exists. The next one is 10; the nonexistence of a finite projective plane of order 10 was proved in 1988 by Lam, Thiel and Swiercz based on an extensive use of computers. Nobody has ever found a projective plane whose order is not a prime power, but the question whether such a plane exists is unsolved.

14.2.3 Suppose that we want to verify the nonexistence of a finite projective plane of order 10 by computer, by simple "brute force": We check that no matter how we specify the appropriate number of subsets of points as lines, one of the conditions (a), (b), or (c) will not hold. How may possibilities do we have to try? About how long would this take?

14.3   Block Designs

The inhabitants of a town like to form clubs. They are socially very sensitive (almost as sensitive as the Fanoans), and don't tolerate any inequalities.

Therefore, they don't allow larger and smaller clubs (because they are afraid that larger clubs might suppress smaller ones). Furthermore, they don't allow some people to be members of more clubs than others, since those who are members of more clubs would have larger influence than the others. Finally, there is one further condition: Each citizen A must behave "equally" toward citizens B and C, A can not be in a tighter relationship with B than C. So A must meet B in the same number of clubs as he/she meets C.

We can formulate these strongly democratic conditions mathematically as follows. The town has v inhabitants; they organize b clubs; every club has the same number of members, say k; everybody belongs to exactly r clubs, and for any pair of citizens, there are exactly λ clubs where both of them are members.

The structure of clubs discussed in the previous paragraphs is called a *block design*. Such a structure consists of a set of v elements, together with a family of k-element subsets of this set (called *blocks*) in such a way that every element occurs in exactly r blocks, and every pair of elements occurs in λ blocks jointly. We denote the number of blocks by b. It is clear that block designs describe what we were discussing when talking about the clubs in the town. In the sequel, sometimes we use this everyday description and sometimes the block design formulation.

Let us see some concrete examples for block designs. One example is given by the Fano plane (Figure 14.1). The points represent the inhabitants of the town, and 3 people form a club if they are on a line.

Let us check that this configuration is indeed a block design: Every club consists of 3 elements (so $k = 3$). Every person belongs to exactly 3 clubs (which means that $r = 3$). Finally, any pair of people belongs to exactly one club by (a) (which means $\lambda = 1$). Thus our configuration is indeed a block design (the number of elements is $v = 7$, the number of blocks is $b = 7$).

The Tictactoe plane gives another block design. Here we have nine points, so $v = 9$; the blocks are the lines, having 3 points on each (so $k = 3$), and there are 12 of them (so $b = 12$). Each point is contained in 4 lines ($r = 4$), and through each pair of points there is a unique line (so $\lambda = 1$).

A block design in which $\lambda \neq 1$ can be obtained from the Cube space if we take the planes as blocks. Clearly, we have $v = 8$ elements; each block contains $k = 4$ elements, the number of blocks is $b = 14$ and every block is contained in $r = 7$ planes. The crucial property is that every pair of points is contained in exactly 3 planes, so we have $\lambda = 3$.

There are some uninteresting, trivial block designs. For $k = 2$, there is only one block design on a given number v of elements: It consists of all $\binom{v}{2}$ pairs. The same construction gives a block design for every $k \geq 2$: We can take all k-subsets as blocks to get a block design with $b = \binom{v}{k}$, $r = \binom{v-1}{k-1}$, and $\lambda = \binom{v-2}{k-2}$. The most boring block design consists of a single block

($k = v$, $b = r = \lambda = 1$). This is so uninteresting that we exclude it from further consideration and don't call it a block design.

Parameters of block designs. Are there any relations among the numbers b, v, r, k, λ? One equation can be derived from the following. Suppose that every club gives a membership card to every one of its members; how many membership cards do they need? There are b clubs and every club has k members, so altogether there are bk membership cards. On the other hand, the town has v inhabitants and everybody has r membership cards, so counting this way we get vr membership cards. In counting the number of membership cards two ways we have to get the same number, so we get

$$bk = vr. \tag{14.1}$$

Let us find another relation. Imagine that the clubs want to strengthen the friendship among their members, so they require that every member should have a dinner one-on-one with each of his/her fellow club members in the clubhouse. How many dinners will a citizen C eat? We can count this in the following two ways:

First reasoning: There are $v - 1$ other citizens in the town, and each of them is in λ clubs jointly with C, so with every one of the other $v - 1$ citizens, C will have to eat λ dinners in the different clubhouses. This means altogether $\lambda(v - 1)$ dinners.

Second reasoning: C is a member in r clubs. Every club has $k - 1$ further members, so in every club C is a member of, C has to eat $k - 1$ dinners. Altogether this means $r(k - 1)$ dinners.

The result of the two counts must be equal:

$$\lambda(v - 1) = r(k - 1). \tag{14.2}$$

14.3.1 If a town has 924 clubs, each club has 21 members, and any 2 persons belong to exactly 2 clubs jointly, then how many inhabitants does the town have? How many clubs does each person belong to? (Don't be surprised: This is a very small town, and everybody belongs to many clubs!)

14.3.2 Show that the assumption that every person is in the same number of clubs is superfluous: It follows from the other assumptions we made about the clubs.

It follows that among the numbers b, v, r, k, λ we can specify at most three freely, and the other two are already determined by the relations (14.1) and (14.2). In fact, we cannot arbitrarily specify even three. Is it possible, for instance, that in a town of 500 people all the clubs have 11 members, and everybody belongs to 7 clubs? The answer is no; please don't read on, but try to prove it yourself (it is not too hard).

Here is our proof: If this were possible, then for the number of clubs b we would get from (14.1) that

$$b = \frac{v \cdot r}{k} = \frac{500 \cdot 7}{11}.$$

But this is not an integer, so these numbers cannot occur.

OK, there is a rather trivial answer to this problem: We must specify three of the given numbers so that in computing the other two using (14.1) and (14.2) we get integer values. But this is not the whole story. There is an important inequality that holds true in every block design, called Fisher's Inequality:

$$b \geq v. \tag{14.3}$$

The proof of this inequality uses mathematical tools that go beyond the scope of this book.

Unfortunately, one can find five numbers b, v, r, k, λ satisfying conditions (14.1), (14.2), and (14.3) for which no block design with these parameters exists. But we are running out of simple, easily checkable necessary conditions. For instance, there is no block design with parameters $b = v = 43$, $k = r = 7$, $\lambda = 1$ (since this block design would be a finite projective plane of order 6, which we know does not exist). These numbers satisfy (14.1), (14.2), and (14.3), and there is no simple way to rule out this block design (just a tedious study of many cases).

14.3.3 (a) Find an example of specific values for v, r, and k where computing b from (14.1) gives an integer value, but (14.2) leads to a contradiction. (b) Find an example of 5 integers b, v, k, r, λ ($b, k, v, r \geq 2$, $\lambda \geq 1$) that satisfy both (14.1) and (14.2), but $b < v$.

14.3.4 For every $v > 1$, construct a block design with $b = v$.

Club badges. In our town, every club has a badge. The town organizes a parade in which everybody participates and everybody is required to wear the badge of one of the clubs where he/she is a member. Can the badges be chosen so that no two persons wear the same badge?

We need, of course, that there are enough different badges, at least as many as citizens. That is, $b \geq v$. This is indeed guaranteed by Fisher's Inequality (14.3). But is this enough? We have to make sure that every citizen is wearing a badge of one of his or her clubs; not just different badges.

The question has some resemblance to the Marriage Theorem (Theorem 10.3.1) described in Chapter 10. To make use of this resemblance, we assign a bipartite graph to our block design (Figure 14.9). The lower set of points represents the people; the upper set of points represents the clubs. We connect point a to point X if citizen a is a member in club X (in Figure

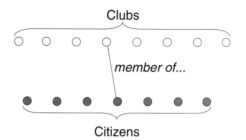

FIGURE 14.9. Representing club membership by a graph

14.9 we have drawn only one edge of the graph). We know the following properties of our graph: from every point below, exactly r edges go up, and from every upper point exactly k edges go down. Below we have v points, above we have b points. If we choose n points from the lower set (obviously, $n \leq v$), then we know that from these n points nr edges leave. Let us denote by m the number of other endpoints of these nr edges.

We claim that $n \leq m$. Since every upper point has degree k, altogether this makes mk edges. The nr edges mentioned above are among these mk edges, and hence

$$nr \leq mk. \tag{14.4}$$

On the other hand, $bk = vr$ by (14.1). By $b \geq v$ we get $k \leq r$, and so

$$mk \leq mr. \tag{14.5}$$

From (14.4) and (14.5) we get

$$nr \leq mr,$$

and therefore $n \leq m$, as claimed.

So any n lower points are connected to at least n upper points. We invoke the Marriage Theorem, more precisely, its version stated in Exercise 10.3.2: We get that there exists a matching of the lower set into the upper set, i.e., there exist v independent edges that connect every lower point to a different point above. This matching tells every citizen which badge to wear.

14.4   Steiner Systems

We have seen that block designs with $k \leq 2$ are trivial; we take a closer look at the next case, $k = 3$. We also take the smallest possible value for λ, namely $\lambda = 1$. Block designs with $k = 3$ and $\lambda = 1$ are called *Steiner systems* (named after Jakob Steiner, a Swiss mathematician in the nineteenth century). The Fano plane and the Tictactoe plane are Steiner systems, but the Cube space is not.

How many inhabitants must a town have to allow a system of clubs that is a Steiner system? In other words, what conditions do we get for v if our block design is a Steiner system? We use equations (14.1) and (14.2), substituting the values $k = 3$ and $\lambda = 1$. We get

$$3b = vr \quad \text{and} \quad 2r = v - 1,$$

and hence

$$r = \frac{v - 1}{2} \tag{14.6}$$

and

$$b = \frac{v(v - 1)}{6}. \tag{14.7}$$

The numbers r and v must be integers, which imposes some conditions on v. Since the denominator in (14.7) is 6 and that in (14.6) is a divisor of 6, the condition imposed concerns the remainder of v upon division by 6. From (14.6) it follows that v must be an odd number, so if we divide it by 6, the remainders can be 1, 3, or 5. This means that v can be written in the forms $6j + 1$, $6j + 3$, or $6j + 5$, where j is an integer. Furthermore, v can not be of the form $6j + 5$, because then by (14.7) we get

$$b = \frac{(6j + 5)(6j + 4)}{6} = 6j^2 + 9j + 3 + \frac{1}{3},$$

which is not an integer.

So v must be of the form $6j + 1$ or $6j + 3$. Taking into consideration that we must have $v > k = 3$, we see that one can have a Steiner system only in towns where the number of inhabitants is $v = 7, 9, 13, 15, 19, 21, \ldots$ etc.

For these numbers one can construct Steiner systems indeed. In the case $v = 7$ we already have seen the Fano plane, and for $v = 9$, the Tictactoe plane. The general construction is quite involved, and we don't describe it here.

14.4.1 Show that for for $v = 7$, the Fano plane is the only Steiner system. (Of course, 7 citizens can set up their clubs in many different ways, by "switching identities." We can think of 7 chairs, with triples forming the clubs specified. The citizens can choose chairs in many different ways.)

14.4.2 Does the Fisher inequality give any further condition on the number of elements in a Steiner system?

Representing the clubs. Imagine that in a town of v people, where the clubs form a Steiner system, people become unhappy about the membership fees, and they create a committee whose task is to protest these high fees. The committee needs at least one member from every club. How many members does this protest committee need to have?

This problem sounds similar to the badge problem discussed at the end of section 14.3, but there are two differences: First, in the badge problem every citizen (his or her interest) was "represented" by one of the clubs the citizen belonged to, while here the clubs will be represented by one of their members; and second (and more significantly), one and the same person can represent several clubs.

Consider citizen Andrew, who does not belong to the committee. Andrew is a member of r clubs, and since the clubs form a Steiner system, we have

$$r = \frac{v-1}{2}$$

(see equation (14.6)).

Every club in which Andrew is a member has two other members, and since Andrew is a member jointly in exactly one club with every other citizen, these $(v-1)/2$ clubs have no members in common other than Andrew. The protest committee must have a member from each of them, and since this member is not Andrew (since he is not a member of the committee), the committee has at least $\frac{v-1}{2}$ members. This means that one needs quite a large committee—almost half of the citizens!

And even this is only a lower bound! Can it be realized, or is there some other argument that gives perhaps an even larger lower bound on the size of the committee? In the case of the Fano plane, our lower bound is $(7-1)/2 = 3$, and indeed, the three points of any line represent every line (any two lines intersect). In the case of the Tictactoe plane, our lower bound is $(9-1)/2 = 4$, but there is no obvious choice of a committee of 4 representing every line; in fact, there is no such choice at all!

This can be seen by a tedious case distinction, but let us offer a nice argument that takes care of many other cases as well. We claim the following surprising fact:

Theorem 14.4.1 *If there exists a committee of size $\frac{v-1}{2}$ representing every club, then this committee itself is also a Steiner system.*

To be more precise, the elements of this new Steiner system will be the v_1 people in the committee, and its blocks will be those clubs for which all three members belong to the committee (such clubs will be called *privileged*).

Proof. To prove that this is indeed a Steiner system, first we show that every two committee members are contained in a privileged club. Suppose not, say Bob and Carl are two committee members who are not jointly contained in a privileged club. This means that the third member of the club to which both Bob and Carl belong (in the original Steiner system) is not on the committee. We may assume that this third member is Andrew. But then in the argument above we see that each club containing Andrew has at least one representative in the committee, and one club has two (Bob and

Carl). This implies that the committee has at least $(v-3)/2+2 = (v+1)/2$ members, a contradiction.

So any two committee members belong to a privileged club. Since no two citizens belong to more than one club in common, no two committee members belong to more than one privileged club in common. So every pair of committee members belongs to exactly one privileged club in common. This proves that the committee is indeed a Steiner system. $\square$

Now, if the 9-element Steiner system could be represented by 4 elements, then we would get a Steiner system on 4 elements, which we already know cannot exist! We get similarly that for Steiner systems on $13, 21, 25, 33, \ldots$ points, more than half of the citizens are needed to represent every club.

14.4.3 Suppose that a Steiner system on v elements contains a subset S of $(v-1)/2$ elements such that those triples of the original system that belong totally to S form a Steiner system. Prove that in this case S forms a representative committee (so every triple of the original Steiner system contains an element of S).

Gender balance. The inhabitants of our town want to set up their clubs so that in addition to forming a Steiner system, they should be "gender-balanced." Ideally, they would like to have the same number of males and females in each club. But realizing that this cannot happen (3 being an odd number), they would settle for less: They require that every club must contain at least one male and at least one female.

In mathematical terms, we have a Steiner system, and we want to color the elements with 2 colors (red and blue, corresponding to "female" and "male") in such a way that no block (club) gets only one color. Let us call such a coloring a *good* 2-coloring of the Steiner system.

Is this possible? Let's start with the first nontrivial special case, the case $v = 7$. We have seen (Exercise 14.4.1) that the only Steiner system in this case is the Fano plane. After a little experimentation we can convince ourselves that there is no way to 2-color this system. In fact, we have stated this already in exercise 14.1.2: If a good 2-coloring were possible, then in any case where the males vote one way and the females the other way, the "line rule" would not provide a clear-cut decision.

But it is not only the Fano plane for which no good 2-coloring is possible:

Theorem 14.4.2 *No Steiner system has a good 2-coloring.*

Proof. After our preparations, this is not hard to prove. Suppose that we have found a good 2-coloring (thus every triple contains differently colored elements). Then the set of red points and the set of blue points each represent all clubs, so (by our discussion above) both sets must contain at least $\frac{v-1}{2}$ elements. Altogether this gives $v-1$ points, so there is only one further point, which is blue or red; say it is red. Then the $(v-1)/2$ blue points form a representative committee, and as we have seen, it is itself a Steiner system. But then any club that is a block in this smaller Steiner system contains only blue points, contradicting the fact that we supposed we had a good coloring.   □

14.4.4 Show that if we allow 3 colors, then both the Fano plane and the Tic-tactoe plane can be colored so that every block gets at least two colors (but not necessarily all three).

The Schoolgirls' Walking Schedule. A teacher has a group consisting of 9 schoolgirls. She takes them for a walk every day; they walk in three lines, with three girls in each line. The teacher wants to arrange the walks so that after several days, every girl should have walked with every other girl in one line exactly once.

14.4.5 How many days do they need for that?

If you've solved the previous exercise, then you know how many days they need, but is it possible to arrange the walk as the teacher wishes? Trying to make a plan from scratch is not easy.

An observation that helps is the following. Call a triple of girls a *block* if they walk in one line at any time. This way we get a Steiner system. We already know a Steiner system on 9 elements, the Tictactoe plane.

Are we done? No, because the problem asks for more than simply a Steiner system: We have to specify which blocks (triples) form lines on each day. The order of the days is clearly irrelevant, so what we need is a splitting of the 12 triples into 4 classes such that each class consists of three disjoint triples (thus giving a walk plan for a day). If we take a careful look at the Tictactoe plane, then we notice that this is exactly how it is constructed: a set of 3 parallel lines gives a walk plan for one day.

Thomas Kirkman, an English amateur mathematician, asked this question about 15 girls (then the girls need 7 days to complete a walking plan). The question for 15 girls remained unsolved for several years, but finally mathematicians found a solution. Obviously, once you have the right plan, it is easy to check the correctness of it, but there are many possible plans to try.

To find a perfect walking plan for the general case when $v = 6j + 3$, instead of 9 or 15, turned out to be a much harder question. It was solved only more than 100 years later, in 1969, by the Indian-American mathematician Ray-Chaudhuri. One should notice that from this result it follows that there exists a Steiner system for every $v = 6j + 3$; even to prove this simpler fact is quite hard.

There is a related question: Suppose that the teacher wants a plan in which every three girls walk together in a line exactly once. It is not hard to see that such a plan would last for $\binom{15}{3}/5 = 91$ days. The triples that walk together form a block design again, but now this is what we called above "trivial" (all triples out of 15 points). So this problem appears easier than the Kirkman Schoolgirl Problem, but its solution (in general, with v schoolgirls walking in lines of k) took even more time: it was solved in 1974 by the Hungarian mathematician Zsolt Baranyai.

14.5   Latin Squares

Look at the little 4×4 tables below. Each of them has the property that on every field we have one of the numbers 0, 1, 2, and 3, in such a way that no number occurs in any row or in any column, more than once. A table with this property is called a (4×4) *Latin square*.

0	1	2	3
1	2	3	0
2	3	0	1
3	0	1	2

0	2	1	3
2	1	3	0
1	3	0	2
3	0	2	1

(14.8)

0	1	2	3
1	0	3	2
2	3	0	1
3	2	1	0

0	1	2	3
3	2	1	0
1	0	3	2
2	3	0	1

(14.9)

It is easy to construct many Latin squares with any number of rows and columns (see Exercise 14.5.2). Once we have a Latin square, it is easy to make many more from it. We can reorder the rows, reorder the columns, or permute the numbers $0, 1, \ldots$ occurring in them. For example, if we replace 1 by 2 and 2 by 1 in the first Latin square in (14.8), we get the second Latin square.

14.5.1 How many 4×4 Latin squares are there? What is the answer if we don't consider two Latin squares different if one of them can be obtained from the other by permuting rows, columns, and numbers?

14.5.2 Construct an $n \times n$ Latin square for every $n > 1$.

Let us have a closer look at the Latin squares in (14.9). If we place these two squares on top of each other, in every field we get an ordered pair of numbers: The first element of the pair comes from the appropriate field of the first square, and the second element from the appropriate field of the second:

0,0	1,1	2,2	3,3
1,3	0,2	3,1	2,0
2,1	3,0	0,3	1,2
3,2	2,3	1,0	0,1

(14.10)

Do you notice something about this composite square? Every field contains a different pair of numbers! From this it follows that each of the possible $4^2 = 16$ pairs occurs exactly once (Pigeonhole Principle). If two Latin squares have this property, we call them *orthogonal*. One may check the orthogonality of two Latin squares in the following way: We take all the fields in the first Latin square that contain 0, and we check the same fields on the second square, to see whether they contain different integers. We do the same with 1, 2, etc. If the squares pass all these checks, then the first square is orthogonal to the second one, and vice versa.

14.5.3 Find two orthogonal 3×3 Latin squares.

Magic squares. If we have two orthogonal Latin squares, we may very easily construct from them a *magic square*. (In a magic square the sums of the numbers in every row and every column are equal.) Consider the pairs in the fields in (14.10). Replace each pair (a, b) by $\overline{ab} = 4a + b$ (in other words, consider $\overline{ab}$ as a two-digit number in base 4). Writing our numbers in decimal notation, we get the magic square shown in (14.11).

0	5	10	15
7	2	13	8
9	12	3	6
14	11	4	1

(14.11)

(This is a magic square indeed: Every row and column sum is 30.) From any two orthogonal Latin squares we can get a magic square using the same method. In every row (and also in every column) the numbers 0, 1, 2, 3 occur exactly once in the first position and exactly once in the second position, so in every row (and column) the sum of the elements is exactly

$$(0 + 1 + 2 + 3) \cdot 4 + (0 + 1 + 2 + 3) = 30,$$

as required in a magic square.

14.5.4 In our magic square we have the numbers 0 through 15, instead of 1 through 16. Try to make a magic square from (14.10) formed by the numbers 1 through 16.

14.5.5 The magic square constructed from our two orthogonal Latin squares is not "perfect", because in a perfect magic square the sums on the diagonals are the same as the row and column sums. From which orthogonal Latin squares can we make perfect magic squares?

Is there a 4×4 Latin square that is orthogonal to both of our Latin squares making up (14.10)? The answer is yes; try to construct it yourself before looking at (14.12). It is interesting to notice that these three Latin squares consist of the same rows, but in different orders.

$$
\begin{array}{|c|c|c|c|}
\hline
0 & 1 & 2 & 3 \\
\hline
2 & 3 & 0 & 1 \\
\hline
1 & 0 & 3 & 2 \\
\hline
3 & 2 & 1 & 0 \\
\hline
\end{array}
\tag{14.12}
$$

14.5.6 Prove that there does not exist a fourth 4×4 Latin square orthogonal to all three Latin squares in (14.9) and (14.12).

14.5.7 Consider the Latin square (14.13). It is almost the same as the previous one in (14.12); but (prove!) there does not exist any Latin square orthogonal to it. So Latin squares that look similar can be very different!

$$
\begin{array}{|c|c|c|c|}
\hline
0 & 1 & 2 & 3 \\
\hline
1 & 3 & 0 & 2 \\
\hline
2 & 0 & 3 & 1 \\
\hline
3 & 2 & 1 & 0 \\
\hline
\end{array}
\tag{14.13}
$$

Latin squares and finite planes. There is a very close connection between Latin squares and finite affine planes. Consider an affine plane of order n; pick any class of parallel lines, and call them "vertical"; pick another class and call them "horizontal". Enumerate the vertical lines arbitrarily, and also the horizontal lines arbitrarily. Thus we can think of the points of the plane as entries of an $n \times n$ table in which every row as well as every column is a line (this is the way we presented the Tictactoe plane at the beginning of this Chapter).

Now consider an arbitrary third parallel class of lines, and again, label the lines arbitrarily $0, 1, \ldots, n - 1$. Each entry of the table (point in the plane) belongs to exactly one line of this third parallel class, and we can write the label of this line in the field. So all the 0 entries will form a line of the plane, all the 1 entries a different, but parallel, line, etc.

Since any two nonparallel lines have exactly one point in common, the line of 0's will meet every row (and similarly every column) exactly once, and the same holds for the lines of 1's, 2's, etc. This implies that the table we constructed is a Latin square.

This is not too exiting so far, since Latin squares are easy to construct. But if we take a fourth parallel class, and construct a Latin square from it, then *these two Latin squares will be orthogonal!* (This is just a translation of the fact that every line from the third parallel class intersects every line from the fourth exactly once.) The affine plane has $n + 1$ parallel classes; two of these were used to set up the table, but the remaining $n - 1$ provide $n - 1$ mutually orthogonal Latin squares.

From the Tictactoe plane we get two orthogonal 3×3 Latin squares this way (not surprisingly, they are just the ones found directly in exercise 14.5.3). From the affine plane of order 5 constructed earlier, we get 4 mutually orthogonal Latin squares, as shown below.

0	1	2	3	4
1	2	3	4	0
2	3	4	0	1
3	4	0	1	2
4	0	1	2	3

0	1	2	3	4
2	3	4	0	1
4	0	1	2	3
1	2	3	4	0
3	4	0	1	2

0	1	2	3	4
3	4	0	1	2
1	2	3	4	0
4	0	1	2	3
0	1	2	3	4

0	1	2	3	4
4	0	1	2	3
3	4	0	1	2
2	3	4	0	1
1	2	3	4	0

This nice connection between Latin squares and affine planes works both ways: If we have $n - 1$ mutually orthogonal Latin squares, we can use them to construct an affine plane in a straightforward way. The points of the plane are the fields in an $n \times n$ table. The lines are the rows and columns, and for every number in $\{0, 1, \ldots, n - 1\}$ and every Latin square, we form a line from those fields that contain this number.

Recall that we have constructed finite planes only of prime order, even though we remarked that they exist for all prime power orders. We can now settle at least the first of the missing orders: Just use this reverse construction to get an affine plane of order 4 from our three mutually orthogonal 4×4 Latin squares in (14.9) and (14.12).

14.6   Codes

We are ready to talk about some *real* applications of the ideas discussed in this Chapter. Suppose that we want to send a message through a noisy channel. The message is (as usual) a long string of bits (0's and 1's), and "noisy" means that some of these bits may be corrupted (changed from 0 to 1 or vice versa). The channel itself could be radio transmission, telephone, internet, or just your compact disc player (in which case the "noise" may be a piece of dirt or a scratch on the disc).

How can we cope with these errors and recover the original message? Of course, a lot depends on the circumstances. Can we ask for a few bits to be resent, if we notice that there is an error? In internet protocols we can; in transmissions from a Mars probe or in listening to music on compact discs, we can't. So in some cases it is enough if we can *detect* whether there is an error in the message we receive, while in other cases we have to be able to *correct* the error just from the received message itself.

The simplest solution is to send the message twice, and check whether any of the bits arrives differently in the two messages (we can repeat each bit immediately, or repeat the whole message; it does not make any difference at this point). This is called a *repetition code*. Certainly, if a bit does not match, then we know something is wrong, but of course we don't know whether the first or second copy of the bit was wrong. So we *detect* an error, but cannot *correct* it. (Of course, it may happen that *both* copies of the bit are corrupted; we have to make the assumption that the channel is not too noisy, so that the probability that this happens is small. We'll come back to this issue.) An easy way to strengthen this is to send the message three times. Then we can also correct the error (for every bit, take as correct the version that arrives at least twice), or in the very noisy situation, at least detect it even if 2 (but not 3) copies of a bit are corrupted.

There is another simple way of detecting errors: a parity check. This is the simple trick of appending a bit to each string of a given length (say, after 7 bits) so that the number of 1's in the extended message is even (so we append a 0 if the number of 1's is already even, and append a 1 if it is odd). The recipient can look at the received block of 8 bits (a byte), and check whether it has an even number of 1's. If so, we consider it OK; otherwise, we know that it contains an error. (Again, in a very noisy channel errors may remain undetected: If two bits of the 8 are changed, then the parity check does not reveal it.)

Here is an example of how a string (namely, 10110010000111) is encoded in these two ways:

110011110000110000000011111   (repetition code)
1011001**0**00001111   (parity check)

These solutions are not cheap; their main cost is that the messages become longer. In the repetition code, the increase is 100%; in the parity check, it is about 14%. If the errors are so rare that we can safely assume that only one bit in (say) every 127 is corrupted, then it suffices to append a parity check bit after every 127 bits, at a cost of less than 1%. (Note that the repetition code can be thought of as inserting a parity check bit after every single bit!)

Is this the best way? To answer this question, we have to make an assumption about the noisiness of the channel. So we assume that we are sending a message of length k, and that there are no more than e errors (corrupted bits). We cannot use all strings of length k to send messages (since then any error would result in another possible message, and the error could not be detected). The set of strings that we use is called a *code*. So a code is a set of 0-1 strings of length k. For $k = 8$ (one byte), the repetition code consists of the following 16 strings:

00000000, 00000011, 00001100, 00001111, 00110000, 00110011,

00111100, 00111111, 11000000, 11000011, 11001100, 11001111,

11110000, 11110011, 11111100, 11111111;

the parity check code consists of all strings of length 8 in which the number of 1's is even (there are $2^7 = 128$ of these, and so we don't list them all here).

We have seen that the parity check code is 1-*error-detecting*, and so is the repetition code. What is the strongest code on 8 bits (detecting the largest number of errors)? The answer is easy: the code consisting of the two codewords

00000000,   11111111

is 7-error-detecting: All 8 bits must be corrupted before we can be fooled. But this code comes with a very high price tag: What it means is that we resend every bit 8 times.

We can construct a more interesting code from the Cube space. This has 8 points, corresponding to the 8 bits. Let us fix an ordering of the points, say $ABCDEFGH$ in Figure 14.7. Every plane P in the geometry will provide a codeword: We send a 1 if the corresponding point lies in the plane P. For example, the bottom-face plane gives the codeword 11110000; the black plane gives the codeword 10100101. We also add the words 00000000 and 11111111, to get a total of 16 codewords.

How good is this code? How many bits must be corrupted before one codeword is changed into another? Assume that the two codewords come from two planes P and Q, which by property (B) of the Cube space are either parallel or intersect each other in a line (i.e., in two points).

First suppose that these planes are parallel. For example, if they are the "black" and "light" planes, then the two codewords they provide are

10100101

01011010.

We wrote the codewords above each other, so that it should be easier to make the following observation: The two planes have no point in common, which (according to the way we constructed the code) means that in no position can both of them have a "1". Since each of them has four 1's, it follows that in no position can both of them have a "0" either. So all 8 bits must be changed before one of them becomes the other.

Second, suppose that the two planes intersect in two points. For example, the "black" plane and the "bottom" plane give the codewords

$$10100101,$$
$$11110000.$$

The two codewords will have two common 1's, and (since each has four 1's) two common 0's. So 4 bits must be changed before one of them becomes the other.

The two further codewords that we added as a kind of an afterthought, all-0 and all-1, are easy to check: We must change 4 bits in them to get a codeword coming from a plane, and 8 bits in them to get one from the other.

What is important from these is that *if we change up to 3 bits in any codeword, we get a string that is not a codeword.* In other words, this code is 3-error-detecting.

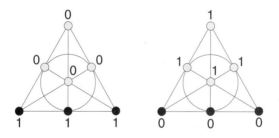

FIGURE 14.10. Two codewords from one line.

The Fano plane provides another interesting code. Again, let each point correspond to a position in the codewords (so the codewords will consist of 7 bits). Each line will provide two codewords, one in which we put 1's for the points on the line and 0's for the points outside, and one in which it is the other way around. Again, we add the all-0 and all-1 strings to get 16 codewords.

Instead of ordering the bits of the codeword, we can think of them as writing 0 or 1 next to each point of the Fano plane. Figure 14.10 illustrates the two codewords associated with a line.

Since we have 16 codewords again, but use only 7 bits, we expect less from these codes than from the codes coming from the Cube space. Indeed, these Fano codes can no longer detect 3 errors. If we start with the codeword defined by a line L (1 on the line and 0 elsewhere), and change the three 1's to 0's, then we get the all-0 string. But it is not only these two special codewords that cause the problem: Again, if we start with the same codeword, and flip the 3 bits on any other line K (from 1 to 0 at the intersection point of K and L, from 0 to 1 at the other two points of K), then we get a codeword coming from a third line (Figure 14.11).

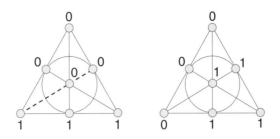

FIGURE 14.11. Three errors are too much for the Fano code: flipping the bits along the dotted line produces another valid codeword.

Going through an argument as above, we can see that

the Fano code is 2-error-detecting: We cannot get a valid codeword from another valid codeword by flipping 1 or 2 bits.

This implies that

the Fano code is 1-error-correcting.

What this means is that not only can we detect if there is an erroneous bit, but we can correct it. Suppose that we receive a string that comes from a valid codeword a by changing a single bit. Could this string come from another valid codeword b? The answer is no: Otherwise, the codewords a and b would differ in only two places, which is not possible.

The codes we derived from the Fano plane and the Cube space are special cases of a larger family of codes called *Reed–Müller codes*. These are very important in practice. For example, the NASA Mariner probes used them to send back images from the Mars. Just as the Cube Code was based on 2-dimensional subspaces of a 3-dimensional space, the code used by the Mariner probes were based on 3-dimensional subspaces of a 5-dimensional space. They worked with blocks of size 32 (instead of 8 as we did), and could correct up to 7 errors in each block. The price was, of course, pretty stiff: There were only 64 codewords used, so to safely transmit 6 bits, one had to package them in 32 bits. But of course, the channel (the space between Earth and Mars) was very noisy!

Error-correcting codes are used all around us. Your CD player uses a more sophisticated error-correcting code (called the Reed–Solomon code) to produce a perfect sound even if the disk is scratched or dusty. Your Internet connection and digital phone use such codes to correct for noise on the line.

14.6.1 Prove that a code is d-error-correcting if and only if it is $(2d)$-error-detecting.

14.6.2 Show that every string of length 7 is either a codeword of the Fano code or arises from a unique codeword by flipping one bit (a code with this property is called *perfect 1-error-correcting*).

Review Exercises

14.6.3 Verify that the Tictactoe plane is the same as the affine plane over the 3-element field.

14.6.4 A lab has 7 employees. Everybody works 3 nights a week: Alice on Monday, Tuesday, and Thursday; Bob on Tuesday, Wednesday, and Friday; etc. Show that any two employees meet exactly one night a week, and for any two nights there is an employee who is working on both. What is the connection with the Fano plane?

14.6.5 The game SMALLSET (which is a simplified version of the commercial card game SET) is played with a deck of 27 cards. Each card has 1, 2, or 3 identical shapes; each shape can be a circle, triangle, or square, and it can be red, blue, or green. There is exactly one card with 2 green triangles, exactly one with 3 red circles, etc. A SET is a triple of cards such that the number of shapes on them is either all the same or all different; the shapes are either all the same or all different; and their colors are all the same or all different. The game consists of putting down 9 cards, face up, and recognizing and removing SETs as quickly as you can; if no SETs are left, 3 new cards are turned up. If no SETs are left and all the remaining cards are turned up, the game is over.

(a) What is the number of SETs?

(b) Show that for any two cards there is exactly one third card that forms with them a SET.

(c) What is the connection between this game and the affine space over the 3-element field?

(d) Prove that at the end of the game, either no cards or at least 6 cards remain.

14.6.6 How many points do the two smallest projective planes have?

14.6.7 Consider the prime field with 13 elements. For every two numbers x and y in the field, consider the triple $\{x + y, 2x + y, 3x + y\}$ of elements of the field. Show that these triples form a block design, and compute its parameters.

14.6.8 Determine whether there exists a block design with the following parameters:

(a) $v = 15$, $k = 4$, $\lambda = 1$;

(b) $v = 8$, $k = 4$, $\lambda = 3$;

(c) $v = 16$, $k = 6$, $\lambda = 1$.

14.6.9 Prove that the Tictactoe plane is the only Steiner system with $v = 9$.

14.6.10 Consider the addition table of the "Days of the Week" number system in Section 6.8. Show that this table is a Latin square. Can you generalize this observation?

14.6.11 Describe the code you get from the projective plane over the 3-element field, analogously to the Fano code. How much error correction/detection does it provide?

15
A Glimpse of Complexity and Cryptography

15.1   A Connecticut Class in King Arthur's Court

In the court of King Arthur[1] there dwelt 150 knights and 150 ladies-in-waiting. The king decided to marry them off, but the trouble was that some pairs hated each other so much that they would not get married, let alone speak! King Arthur tried several times to pair them off but each time he ran into conflicts. So he summoned Merlin the Wizard and ordered him to find a pairing in which every pair was willing to marry. Now, Merlin had supernatural powers, and he saw immediately that none of the 150! possible pairings was feasible, and this he told the king. But Merlin was not only a great wizard, but a suspicious character as well, and King Arthur did not quite trust him. "Find a pairing or I shall sentence you to be imprisoned in a cave forever!" said Arthur.

Fortunately for Merlin, he could use his supernatural powers to browse forthcoming scientific literature, and he found several papers in the early twentieth century that gave the *reason* why such a pairing could not exist. He went back to the king when all the knights and ladies were present, and asked a certain 56 ladies to stand on one side of the king and 95 knights on the other side, and asked, "Is any one of you ladies willing to marry any of

---

[1]From L. Lovász and M.D. Plummer: *Matching Theory*, Akadémiai Kiadó, North Holland, Budapest, 1986 (with slight modifications), with the kind permission of Mike Plummer. The material was developed as a handout at Yale University, New Haven, Connecticut.

these knights?" and when all said "No!" Merlin said, "Oh King, how can you command me to find a husband for each of these 56 ladies among the remaining 55 knights?" So the king, whose courtly education did include the Pigeonhole Principle, saw that in this case Merlin had spoken the truth, and he graciously dismissed him.

Some time elapsed and the king noticed that at the dinners served for the 150 knights at the famous round table, neighbors often quarreled and even fought. Arthur found this bad for the digestion and so once again he summoned Merlin and ordered him to find a way to seat the 150 knights around the table so that each of them should sit between two friends. Again, using his supernatural powers Merlin saw immediately that none of the 150! seatings would do, and this he reported to the king. Again, the king bade him find one or explain why it was impossible. "Oh, I wish there were some simple reason I could give to you! With some luck there could be a knight having only one friend, and so you, too, could see immediately that what you demand from me is impossible. But alas!, there is no such simple reason here, and I cannot explain to you mortals why no such seating exists, unless you are ready to spend the rest of your life listening to my arguments!" The king was naturally unwilling to do that, and so Merlin has lived imprisoned in a cave ever since. (A severe loss for applied mathematics!)

The moral of this tale is that there are properties of graphs that when they hold, are easily proven to hold. If a graph has a perfect matching, or a Hamilton cycle, this can be "proved" easily by exhibiting one. If a bipartite graph does *not* have a perfect matching, then this can be "proved" by exhibiting a subset X of one color class that has fewer than $|X|$ neighbors in the other. The reader (and King Arthur!) are directed to Figure 15.1, in which the graph on the left-hand side has a perfect matching (indicated by the heavy lines), but the graph on the right-hand side does not. To convince ourselves (and the king) of the latter, consider the four black points and their neighbors.

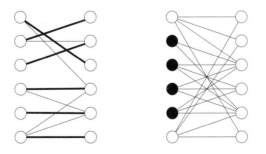

FIGURE 15.1. A bigraph with a perfect matching and one without

Most graph-theoretic properties that interest us have this logical structure. If it is easy to prove (certify, exhibit) that the property holds, then the property is called (in the jargon of computer science) an *NP property* (if you really want to know, NP is the abbreviation of *Nondeterministic Polynomial Time*, but it would be difficult to explain where this highly technical phrase comes from). The two problems that Merlin had to face—the existence of a perfect matching and the existence of a Hamilton cycle—are clearly NP properties. But NP properties also appear quite frequently in other parts of mathematics. A very important NP property of natural numbers is their *compositeness*: If a natural number is composite, then this can be exhibited easily by showing a decomposition $n = ab$ $(a, b > 1)$.

The remarks we have made so far explain how Merlin can remain free if he is lucky and the task assigned to him by King Arthur has a solution. For instance, suppose he could find a good way to seat the knights for dinner. He could then convince King Arthur that his seating plan was "good" by asking if there was anybody sitting beside any enemy of his (or just wait and see if the dinner was quiet). This shows that the property of the corresponding "graph of friendships" that it contains a Hamilton cycle is an NP property. But how could he survive Arthur's wrath in the case of the marriage problem and not in the case of the seating problem when these problems do *not* have solutions? What distinguishes the nonexistence of a Hamilton cycle in a graph from the nonexistence of a perfect matching in a bipartite graph? From our tale, we hope the answer is clear: *the nonexistence of a perfect matching in a bipartite graph is also an NP property* (this is a main implication of the Marriage Theorem, Theorem 10.3.1), while the nonexistence of a Hamilton cycle in a graph is not! (To be precise, no proof of this latter fact is known, but there is strong evidence in favor of it.)

So for certain NP properties the negation of the property is again an NP property. A theorem asserting the equivalence of an NP property with the negation of another NP property is called a *good characterization*. There are famous good characterizations throughout graph theory and elsewhere.

Many NP properties are even better. Facing the problem of marrying off his knights and ladies, Arthur himself (say, after reading this book) could decide himself whether or not it was solvable: He could run the algorithm described in Section 10.4. A lot of work, but probably doable with the help of quite ordinary people, without using the supernatural talents of Merlin. Properties that can be decided efficiently are called properties *in the class P* (here P stand for *Polynomial Time*, an exact but quite technical definition of the phrase "efficiently"). Many other simple properties of graphs discussed in this book also belong to this class, such as connectivity and the existence of a cycle. One of our favorite problems, that of deciding whether a number is prime, was shown to be in this class just before our book went to press. (The algorithm we described in Section 6.10 does not quite qualify for the class P, because it uses a random selection of the base a.)

The introduction of the notions of Polynomial Time and NP properties signaled the birth of modern complexity theory. Notions and paradigms from this theory have penetrated a large part of mathematics and its applications. In the sequel we describe how ideas of complexity theory can be applied in one of the most important areas of theoretical computer science, namely, cryptography.

15.2   Classical Cryptography

Ever since writing was invented, people have been interested not only in using it to communicate with their partners, but also in trying to conceal the content of their messages from their adversaries. This leads to cryptography (or cryptology), the science of secret communication.

The basic situation is that one party, say King Arthur, wants to send a message to King Bela. There is, however, a danger that the evil Caesar Caligula intercepts the message and learns things that he is not supposed to know about. The message, understandable even for Caligula, is called the *plain text*. To protect its content, King Arthur *encrypts* his message. When King Bela receives it, he must *decrypt* it in order to be able to read it. For the kings to be able to encrypt and decrypt the message, they must know something that Caligula does not know: this information is the *key*.

Many cryptosystems have been used in history; most of them, in fact, turn out to be insecure, especially if the adversary can use powerful computing tools to break it.

Perhaps the simplest method is *substitution code*: We replace each letter of the alphabet by another letter. The *key* is the table that contains for each letter the letter to be substituted for it. While a message encrypted this way looks totally scrambled, substitution codes are, in fact, easy to break. Solving Exercise 15.2.1 will make it clear how the length and positions of the words can be used to figure out the original meaning of letters if the word breaks are preserved (i.e., "space" is not replaced by another character). But even if the splitting into words is hidden, an analysis of the frequency of various letters provides enough information to break the substitution code.

One-time pads. There is another simple frequently used method that is much more secure: the use of "one-time pads." This method is very safe; it was used during World War II for communication between the American president and the British prime minister. Its disadvantage is that it requires a very long key ("pad"), which can only be used once.

A one-time pad is a randomly generated string of 0's and 1's. Here is one:

110001110000100001100101001001001011001100101011000011101101000010

Both Kings Arthur and Bela know this sequence (it was sent well in advance by a messenger). Now King Arthur wants to send the following message to King Bela:

ATTACK MONDAY

First, he has to convert it to 0's and 1's. It is not clear that medieval kings had the knowledge to do so, but the reader should be able to think of various ways: using ASCII codes, or Unicodes of the letters, for example. But we want to keep things simple, so we just number the letters from 1 to 26, and then write down the binary representation of the numbers, putting 0's in front so that we get a string of length 5 for each letter. Thus we have "00001" for A, "00010" for B, etc. We use "00000" for "space." The above message becomes

00001100101001000001000110101100000011010111101110001000000111001

This might look cryptic enough, but Caligula (or rather one of the excellent Greek scientists he keeps in his court as slaves) could easily figure out what it stands for. To encrypt it, Arthur adds the one-time pad to the message bit-by-bit. To the first bit of the message (which is 0) he adds the first bit of the pad (1) and writes down the first bit of the encoded message: $0 \oplus 1 = 1$. He computes the second, third, etc., bits similarly: $0 \oplus 1 = 1$, $0 \oplus 0 = 0$, $0 \oplus 0 = 0$, $1 \oplus 0 = 1$, $1 \oplus 1 = 0, \ldots$. (Note that he uses this strange addition from the 2-element field, for which $1 \oplus 1 = 0$.) Another way of saying what King Arthur does is the following: if the k-th bit of the pad is 1, he flips the k-th bit of the text; else, he leaves it as it was.

So Arthur computes the encoded message:

11001011101011000111010010001000101111100101000010000110110111011

He sends this to King Bela, who, using the one-time pad, can easily flip back the appropriate bits, and recover the original message.

But Caligula does not know the one-time pad (nor do his excellent scientists), so he has no idea about which bits were flipped, and so he is helpless. The message is safe.

It can be expensive to make sure that the sender and the receiver both have such a common key; but note that the key can be sent at a safer time and by a completely different method than the message (moreover, it may be possible to agree on a key even without actually passing it; but this would lead us too far into cryptography).

Once the Kings managed to pass the key to each other, it is tempting to reuse it; after all, if Bela encrypts his reply by the same pad, it is still completely random-looking. In solving exercises 15.2.2 and 15.2.3 you will see that this is not a good idea, along with some other weaknesses of the one-time pad method.

15.2.1 For the following message, the kings used substitution code. Caligula intercepted the message and quite easily broke it. Can you do it too?

U GXUAY LS ZXMEKW AMG TGGTIY HMD TAMGXSD LSSY,
FEG GXSA LUGX HEKK HMDIS. FSKT

15.2.2 At one time, Arthur made the mistake of using the one-time pad shifted: The first bit of the plain text he encoded using the second bit of the pad, the second bit of the plain text he encoded using the third bit of the pad etc. He noticed his error after he sent the message off. Being afraid that Bela would not understand his message, he encoded it again (now correctly) using the same one-time pad, and sent it to Bela by another courier, explaining what happened.

Caligula intercepted both messages, and was able to recover the plain text. How?

15.2.3 The Kings were running low on one-time pads, and so Bela had to use the same pad to encode his reply as they used for Arthur's message. Caligula intercepted both messages, and was able to reconstruct both plain texts. How?

15.3   How to Save the Last Move in Chess

Modern cryptography started in the late 1970's with the idea that it is not only lack of information that can protect our message against an unauthorized eavesdropper, but also the *computational complexity* of processing it. The idea can be illustrated by the following simple example.

Alice and Bob are playing chess over the phone. They want to interrupt the game for the night; how can they do it so that the person to move should not get the improper advantage of being able to think about his move the whole night? At a tournament, the last move is not made on the board, only written down, put in an envelope, and deposited with the referee. The next morning, the envelope is opened, and the other player learns what the move was as his clock begins to run. But now the two players have no referee, no envelope, no contact other than the telephone line. The player making the last move (say, Alice) has to send Bob some message. The next morning (or whenever they continue the game) she has to give some additional information, some "key," which allows Bob to reconstruct the move. Bob should not be able to reconstruct Alice's move without the key; Alice should not be able to change her mind overnight and modify her move.

Surely, this seems to be impossible! If she gives enough information the first time to uniquely determine her move, Bob will know the move too soon; if the information given the first time allows several moves, then she can think about it overnight, figure out the best among these, and give the remaining information, the "key," accordingly.

If we measure information in the sense of classical information theory, then there is no way out of this dilemma. But complexity comes to our help: it is not enough to *communicate* information, it must also be *processed*.

So here is a solution to the problem, using elementary number theory! (Many other schemes can be designed.) Alice and Bob agree to encode every move as a 4-digit number (say, '11' means 'K', '6' means 'f', and '3' means itself, so '1163' means 'Kf3'). So far, this is just notation.

Next, Alice extends the four digits describing her move to a prime number $p = 1163\ldots$ with 200 digits. She also generates another prime q with 201 digits and computes the product $N = pq$ (this would take rather long on paper, but is trivial using a personal computer). The result is a number with 400 or 401 digits; she sends this number to Bob.

Next morning, she sends both prime factors p and q to Bob. He reconstructs Alice's move from the first four digits of the smaller prime. To make sure that Alice was not cheating, he should check that p and q are primes and that their product is N.

Let us argue that this protocol does the job.

First, Alice cannot change her mind overnight. This is because the number N contains all the information about her move: This is encoded as the first four digits of the smaller prime factor of N. So Alice commits herself to the move when sending N.

But exactly because the number N contains all the information about Alice's move, Bob seems to have the advantage, and he indeed would have if he had unlimited time or unbelievably powerful computers. What he has to do is to find the prime factors of the number N. But since N has 400 digits (or more), this is a hopelessly difficult task with current technology.

Can Alice cheat by sending a different pair (p', q') of primes the next morning? No, because Bob can easily compute the product $p'q'$, and check that this is indeed the number N that was sent the previous night. (Note the role of the *uniqueness* of prime factorization, Theorem 6.3.1.)

All the information about Alice's move is encoded in the first 4 digits of the smaller prime factor p. We could say that the rest of p and the other prime factor q serve as a "deposit box": This box hides this information from Bob, and it can be opened only if the appropriate key (the factorization of N) is available. The crucial ingredient of this scheme is *complexity*: the computational difficulty to find the factorization of an integer.

With the spread of electronic communication in business, many solutions of traditional correspondence and trade must be replaced by electronic versions. We have seen an electronic "deposit box" above. Other schemes (similar or more involved) can be found for electronic passwords, authorization, authentication, signatures, watermarking, etc. These schemes are extremely important in computer security, cryptography, automatic teller machines, and many other fields. The protocols are often based on simple number theory; in the next section we discuss (a very simplified version of) one of them.

15.3.1 Motivated by the one-time pad method, Alice suggests the following protocol for saving the last move in their chess game: In the evening, she encrypts her move (perhaps with other text added, to make it reasonably long) using a randomly generated 0-1 sequence as the key (just like in the one-time pad method). The next morning she sends the key to Bob, so that he can decrypt the message. Should Bob accept this suggestion?

15.3.2 Alice modifies her suggestion as follows: instead of the random 0-1 sequence, she offers to use a random, but meaningful text as the key. For whom would this be advantageous?

15.4   How to Verify a Password—Without Learning it

In a bank, a cash machine works by name and password. This system is safe as long as the password is kept secret. But there is one weak point: The computer of the bank must store the password, and the administrator of this computer may learn it and later misuse it.

Complexity theory provides a scheme whereby the bank can verify that the customer does indeed know the password—without storing the password itself! At first glance this looks impossible—just as the problem with filing the last chess move was. And the solution (at least the one we discuss here) uses the same kind of construction as our telephone chess example.

Suppose that the password is a 100-digit prime number p (this is, of course, too long for everyday use, but it illustrates the idea best). When the customer chooses the password, he chooses another prime q with 101 digits, forms the product $N = pq$ of the two primes, and tells the bank the number N. When the teller machine is used, the customer tells his name and the password p. The computer of the bank checks whether or not p is a divisor of N; if so, it accepts p as a proper password. The division of a 200 digit number by a 100 digit number is a trivial task for a computer.

Let us assume that the system administrator learns the number N stored along with the files of our customer. To use this in order to impersonate the customer, he has to find a 100-digit number that is a divisor of N; but this is essentially the same problem as finding the prime factorization of N, and this is hopelessly difficult. So—even though all the necessary information is contained in the number N—the computational complexity of the factoring problem protects the password of the customer!

15.5   How to Find These Primes

In our two simple examples of "modern cryptography," as well as in almost all the others, one needs large prime numbers. We know that there are arbitrarily large primes (Theorem 6.4.1), but are there any with 200 digits, starting with 1163 (or any other 4 given digits)? Maple found (in a few seconds on a laptop!) the smallest such prime number:

1163000000000000000000000000000000000000000000000000
0000000000000000000000000000000000000000000000000000
0000000000000000000000000000000000000000000000000000
0000000000000000000000000000000000000000000000000371

The smallest 200 digit integer starting with 1163 is $1163 \cdot 10^{196}$. This is of course not a prime, but above we found a prime very close by. There must be zillions of such primes! Indeed, a computation very similar to what we did in section 6.4 shows that the number of primes Alice can choose from is about $1.95 \cdot 10^{193}$.

This is a large number of possibilities, but how do we find one? It would not be good to use the prime above (the smallest eligible): Bob could guess this and thereby find out Alice's move. What Alice can do is to fill in the missing 196 digits randomly, and then test whether the number she obtains is a prime. If not, she can throw it away and try again. As we computed in Section 6.4, one in every 460 numbers with 200 digits is a prime, so on the average in 460 trials she gets a prime. This looks like a lot of trials, but of course she uses a computer; here is one we computed for you with this method (in a few seconds again):

11631467128765557632799097045596606908283654760066688738144893546624743604198911046804111038868958805745715572480009569639174033385458418593535488622323782317577559864739652701127177097278389465414589

So we see that in the "envelope" scheme above, both computational facts mentioned in Section 6.10 play a crucial role: It is easy to test whether a number is a prime (and thereby it is easy to compute the encryption), but it is difficult to find the prime factors of a composite number (and so it is difficult to break the cryptosystem).

15.6 Public Key Cryptography

Cryptographic systems used in real life are more complex than those described in the previous section; but they are based on similar principles. In this section we sketch the math behind the most commonly used system, the RSA code (named after its inventors, Rivest, Shamir, and Adleman).

The protocol. Alice generates two 100-digit prime numbers, p and q and computes their product $m = pq$. Then she generates two 200-digit numbers d and e such that $(p-1)(q-1)$ is a divisor $ed-1$. (We will return to the question of how this is done.)

The numbers m and e she publishes on her web site (or in the phone book), but the prime factors p and q and the number d remain her closely guarded secrets. The number d is called her *private key*, and the number

e, her *public key* (the numbers p and q she may even forget; they will not be needed to operate the system, just to set it up).

Suppose that Bob wants to send a message to Alice. He writes the message as a number x (we have seen before how to do this). This number x must be a non-negative integer less than m (if the message is longer, he can just break it up into smaller chunks).

The next step is the trickiest: Bob computes the remainder of x^e modulo m. Since both x and e are huge integers (200 digits), the number x^d has more than 10^{200} digits; we could not even write it down, let alone compute it! Luckily, we don't have to compute this number, only its remainder when dividing by m. This is still a large number, but at least it can be written down in 2 or 3 lines.

So let r be this remainder; this is sent to Alice. When she receives it, she can decrypt it using her private key d by carrying out essentially the same procedure as Bob did: She computes the remainder of r^d modulo m. And—a black magic of number theory, until you see the explanations—this remainder is just the plain text x.

What if Alice wants to send a message to Bob? He also needs to go through the trouble of generating his private and public keys. He has to pick two primes p' and q', compute their product m', select two positive integers d' and e' such that $(p'-1)(q'-1)$ is a divisor of $e'd'-1$, and finally publish m' and e'. Then Alice can send him a secure message.

The black math magic behind the protocol. The key fact from mathematics we use is Fermat's Theorem, Theorem 6.5.1. Recall that x is the plain text (written as an integer) and the encrypted message r is the remainder of x^e modulo m. So we can write

$$r \equiv x^e \pmod{m}.$$

To decrypt, Alice raises this to the dth power to get

$$r^d \equiv x^{ed} \pmod{m}.$$

To be more precise, Alice computes the remainder x' of r^d modulo m, which is the same as the remainder of x^{ed} modulo m. We want to show that this remainder is precisely x. Since $0 \le x < m$, it suffices to argue that $x^{ed} - x$ is divisible by m. Since $m = pq$ is the product of two distinct primes, it suffices to prove that $x^{ed} - x$ is divisible by both p and q.

Let us consider divisibility by p, for example. The main property of e and d is that $ed - 1$ is divisible by $(p-1)(q-1)$, and hence also by $p-1$. This means that we can write $ed = (p-1)l+1$, where l is a positive integer. We have

$$x^{ed} - x = x\big(x^{(p-1)l} - 1\big).$$

Here $x^{(p-1)l} - 1$ is divisible by $x^{p-1} - 1$ (see Exercise 6.1.6), and so $x\big(x^{(p-1)l} - 1\big)$ is divisible by $x^p - x$, which in turn is divisible by p by Fermat's Theorem.

How to do all this computation. We already discussed how to find primes, and Alice can follow the method described in section 6.10.

The next issue is the computation of the two keys e and d. One of them, say e, Alice can choose at random, from the range $1, \ldots, (p-1)(q-1)-1$. She has to check that it is relatively prime to $(p-1)(q-1)$; this can be done efficiently with the help of the Euclidean Algorithm discussed in Section 6.6. If the number she chose is not relatively prime to $(p-1)(q-1)$, she just throws it out and tries another one. This is similar to the method we used to find a prime, and it is not hard to see that she will find a good number in about the same number of trials as it takes to find a prime.

But if she finally succeeds and the Euclidean Algorithm shows that she found a number e relatively prime to $(p-1)(q-1)$, then (as in Section 6.6) it also gives two integers u and v such that

$$eu - (p-1)(q-1)v = 1.$$

So $eu - 1$ is divisible by $(p-1)(q-1)$. Let d denote the remainder of u modulo $(p-1)(q-1)$, then $ed - 1$ is also divisible by $(p-1)(q-1)$, and so we have found a suitable private key d.

Finally, how do we compute the remainder of x^e modulo m, when just to write down x^e would fill the universe? This is done in the same way as described in Section 6.10.

Signatures, etc. There are many other useful things this system can do. For example, suppose that Alice gets a message from Bob as described above. How can she know that it indeed came from Bob? Just because it is signed "Bob," it could have come from anybody. But Bob can do the following. First, he encrypts the message with his private key, then adds "Bob," and encrypts it again with Alice's public key. When Alice receives it, she can decrypt it with her private key. She'll see a still encrypted message, signed "Bob." She can cut away the signature, look up Bob's public key in the phonebook, and use it to decrypt the message.

Could anyone have faked this message? No, because the interloper would have to use Bob's private key to encrypt the message (using any other key would mean that Alice gets garbage when she decrypts the message by Bob's public key, and she would know immediately that it is a fake).

One can use similar tricks to implement many other electronic gadgets, using the RSA public key system: authentication, watermarking, etc.

Security. The security of the RSA protocol is a difficult issue, and since its inception in 1977, thousands of researchers have investigated it. The fact that no attack has been generally successful is a good sign; but unfortunately, no exact proof of its security has been found (and it appears that current mathematics lacks the tools to provide such a proof).

We can give, however, at least some arguments that support its security. Suppose that you intercept the message of Bob, and want to decipher it.

You know the remainder r (this is the intercepted message). You also know Alice's public key e, and the number m. One could think of two lines of attack: Either you can figure out her private key d and then decrypt the message just as she does, or you could somehow more directly find the integer x, knowing the remainder of x^e modulo m.

Unfortunately, there is no theorem stating that either of this is impossible in less than astronomical time. But one can justify the security of the system with the following fact: *if one can break the RSA system, then one can use the same algorithm to find the prime factors of m* (see Exercise 15.6.1). The factorization problem has been studied by many people, and no efficient method has been found, which makes the security of RSA quite probable.

15.6.1 Suppose that Bob develops an algorithm that can break RSA in the first, more direct, way described above: Knowing Alice's public key m and e, he can find her private key d.

(a) Show that he can use this to find the number $(p-1)(q-1)$;

(b) from this, he can find the prime factorization $m = pq$.

The real world. How practical could such a complicated system be? It seems that only a few mathematicians could ever use it. But in fact you have probably used it yourself hundreds of times! RSA is used in SSL (Secure Socket Layer), which in turn is used in https (secure http). Any time you visit a "secure site" on the Internet (to read your e-mail or to order merchandise), your computer generates a public and private key for you, and uses them to make sure that your credit card number and other personal data remain secret. It does not have to involve you in this at all; all you notice is that the connection is a bit slower.

In practice, the two 100-digit primes are not considered sufficiently secure. Commercial applications use more than twice this length, military applications, more than 4 times.

While the hairy computations of raising the plain text x to an exponent that itself has hundreds of digits are surprisingly efficient, it would still be too slow to encrypt and decrypt each message this way. A way out is to send, as a first message, the key to a simpler encryption system (think of a one-time pad, although one uses a more efficient system in practice, like DES, the Digital Encryption Standard). This key is then used for a few minutes to encode the messages going back and force, then thrown away. The idea is that in a short session, the number of encoded messages is not enough for an eavesdropper to break the system.

16
Answers to Exercises

1 Let's Count!

1.1 A Party

1.1.1. $7 \cdot 6 \cdots 2 \cdot 1 = 5040$.

1.1.2. Carl: $15 \cdot 2^3 = 120$. Diane: $15 \cdot 3 \cdot 2 \cdot 1 = 90$.

1.1.3. Bob: $9 \cdot 7 \cdot 5 \cdot 3 = 945$. Carl: $945 \cdot 2^5 = 302,40$. Diane: $945 \cdot 5 \cdot 4 \cdot 3 \cdot 2 \cdot 1 = 113,400$.

1.2 Sets and the Like

1.2.1. (a) all houses in a street; (b) an Olympic team; (c) class of '99; (d) all trees in a forest; (e) the set of rational numbers; (f) a circle in the plane.

1.2.2. (a) soldiers; (b) people; (c) books; (d) animals.

1.2.3. (a) all cards in a deck; (b) all spades in a deck; (c) a deck of Swiss cards; (d) nonnegative integers with at most two digits; (e) nonnegative integers with exactly two digits; (f) inhabitants of Budapest, Hungary.

1.2.4. Alice, and the set whose only element is the number 1.

1.2.5. No.

1.2.6. $\emptyset, \{0\}, \{1\}, \{3\}, \{0,1\}, \{0,3\}, \{1,3\}, \{0,1,3\}$. 8 subsets.

1.2.7. women; people at the party; students of Yale.

1.2.8. $\{a\}, \{a, c\}, \{a, d\}, \{a, e\}, \{a, c, d\}, \{a, c, e\}, \{a, d, e\}, \{a, c, d, e\}$.

1.2.9. $\mathbb{Z}$ or $\mathbb{Z}_+$. The smallest is $\{0, 1, 3, 4, 5\}$.

1.2.10. (a) $\{a, b, c, d, e\}$. (b) The union operation is associative. (c) The union of any set of sets consists of those elements that are elements of at least one of the sets.

1.2.11. The union of a set of sets $\{A_1, A_2, \ldots, A_k\}$ is the smallest set containing each A_i as a subset.

1.2.12. $6, 9, 10, 14$.

1.2.13. The cardinality of the union is at least the larger of n and m and at most $n + m$.

1.2.14. (a) $\{1, 3\}$; (b) $\emptyset$; (c) $\{2\}$.

1.2.15. The cardinality of the intersection is at most the minimum of n and m.

1.2.16. Commutativity (1.2) is obvious. To show that $(A \cap B) \cap C = A \cap (B \cap C)$, it suffices to check that both sides consist of those elements that belong to all three of A, B, and C. The proof of the other identity in (1.3) is similar. Finally, one can prove (1.4) completely analogously to the proof of (1.1).

1.2.17. The common elements of A and B are counted twice on both sides; the elements in either A or B but not both are counted once on both sides.

1.2.18. (a) The set of negative even integers and positive odd integers. (b) B.

1.3 The Number of Subsets

1.3.1. (a) Powers of 2. (b) $2^n - 1$. (c) sets not containing the last element.

1.3.2. 2^{n-1}.

1.3.3. Divide all subsets into pairs such that each pair differs only in their first element. Each pair contains an even and an odd subset, so their numbers are the same.

1.3.4. (a) $2 \cdot 10^n - 1$; (b) $2 \cdot (10^n - 10^{n-1})$.

1.4.1. 101.

1.4.2. $1 + \lfloor n \lg 2 \rfloor$.

1.5 Sequences

1.5.1. The trees have 9 and 12 leaves, respectively.

1.5.2. $5 \cdot 4 \cdot 3 = 60$.

1.5.3. 3^{13}.

1.5.4. $6 \cdot 6 = 36$.

1.5.5. 12^{20}.

1.5.6. $(2^{20})^{12}$.

1.6 Permutations

1.6.1. $n!$.

1.6.2. (a) $7 \cdot 5 \cdot 3 \cdot 1 = 105$. (b) $(2n - 1) \cdot (2n - 3) \cdots 3 \cdot 1$.

1.7 The Number of Ordered Subsets

1.7.1. (We don't think you could really draw the whole tree; it has almost 10^{20} leaves. It has 11 levels of nodes.)

1.7.2. (a) $100!$. (b) $90!$. (c) $100!/90! = 100 \cdot 99 \cdots 91$.

1.7.3. $\frac{n!}{(n-k)!} = n(n-1) \cdot (n-k+1)$.

1.7.4. In one case, repetition is not allowed, while in the other case, it is allowed.

1.8 The Number of Subsets of a Given Size

1.8.1. Handshakes; lottery; hands in bridge.

1.8.2. See Pascal's Triangle in Chapter 3.

1.8.3. $\binom{n}{0} = \binom{n}{n} = 1$, $\binom{n}{1} = \binom{n}{n-1} = n$.

1.8.4. An algebraic proof of (1.7) is straightforward. In (1.8), the right-hand side counts k-subsets of an n-element set by separately counting those that contain a given element and those that do not.

1.8.5. An algebraic proof is easy. A combinatorial interpretation: n^2 is the number of all ordered pairs (a, b) with $a, b \in \{1, 2, \ldots, n\}$, and $\binom{n}{2}$ is the number of ordered pairs (a, b) among these with $a < b$ (why?). To count the remaining ordered pairs (a, b) (those with $a \geq b$), add 1 to their first entry. Then we get a pair (a', b) with $1 \leq a', b \leq n + 1$, $a' > b$, and conversely, every such pair is obtained this way. Hence the number of these pairs is $\binom{n+1}{2}$.

1.8.6. Again, an algebraic proof is easy. A combinatorial interpretation: We can choose a k-element set by first choosing one element (n possibilities) and then choosing a $(k-1)$-element subset of the remaining $n-1$ elements ($\binom{n-1}{k-1}$ possibilities). But we get every k-element subset exactly k times (depending on which of its elements was chosen first), so we have to divide the result by k.

1.8.7. Both sides count the number of ways to divide an a-element set into three sets with $a - b$, $b - c$, and c elements.

2 Combinatorial Tools

2.1 Induction

2.1.1. One of n and $n + 1$ is even, so the product $n(n + 1)$ is even. By induction: true for $n = 1$; if $n > 1$, then $n(n + 1) = (n - 1)n + 2n$, and $n(n - 1)$ is even by the induction hypothesis, $2n$ is even, and the sum of two even numbers is even.

2.1.2. True for $n = 1$. If $n > 1$, then

$$1 + 2 + \cdots + n = (1 + 2 + \cdots + (n - 1)) + n = \frac{(n - 1)n}{2} + n = \frac{n(n + 1)}{2}.$$

2.1.3. The youngest person will count n handshakes. The 7th oldest will count 6 handshakes. So they count $1 + 2 + \cdots + n$ handshakes. We already know that there are $n(n + 1)/2$ handshakes.

2.1.4. Compute the area of the rectangle in two different ways.

2.1.5. By induction on n. True for $n = 2$. For $n > 2$, we have

$$1 \cdot 2 + 2 \cdot 3 + 3 \cdot 4 + \cdots + (n - 1) \cdot n = \frac{(n - 2) \cdot (n - 1) \cdot n}{3} + (n - 1) \cdot n$$
$$= \frac{(n - 1) \cdot n \cdot (n + 1)}{3}.$$

2.1.6. If n is even, then $1 + n = 2 + (n - 1) = \cdots = \left(\frac{n}{2} - 1\right) + \frac{n}{2} = n + 1$, so the sum is $\frac{n}{2}(n + 1) = \frac{n(n+1)}{2}$. If n is odd, then we have to add the middle term separately.

2.1.7. If n is even, then $1 + (2n - 1) = 3 + (2n - 3) = \cdots = (n - 1) + (n + 1) = 2n$, so the sum is $\frac{n}{2}(2n) = n^2$. Again, if n is odd, then the solution is similar, but we have to add the middle term separately.

2.1.8. By induction. True for $n = 1$. If $n > 1$, then

$$1^2 + 2^2 + \cdots + (n - 1)^2 = \left(1^2 + 2^2 + \cdots + (n - 1)^2\right) + n^2$$
$$= \frac{(n - 1)n(2n - 1)}{6} + n^2 = \frac{n(n + 1)(2n + 1)}{6}.$$

2.1.9. By induction. True for $n = 1$. If $n > 1$ then

$$2^0 + 2^1 + 2^2 + \cdots + 2^{n-1} = \left(2^0 + 2^1 + \cdots + 2^{n-2}\right) + 2^{n-1}$$
$$= (2^{n-1} - 1) + 2^{n-1} = 2^n - 1.$$

2.1.10. (Strings) True for $n = 1$. If $n > 1$ then to get a string of length n we can start with a string of length $n - 1$ (this can be chosen in k^{n-1} ways by the induction hypothesis) and append an element (this can be chosen in k ways). So we get $k^{n-1} \cdot k = k^n$.

(Permutations) True for $n = 1$. To seat n people, we can start with seating the oldest (this can be done in n ways) and then seating the rest (this can be done in $(n-1)!$ ways by the induction hypothesis). We get $n \cdot (n-1)! = n!$.

2.1.11. True if $n = 1$. Let $n > 1$. The number of handshakes between n people is the number of handshakes by the oldest person (this is $n - 1$) plus the number of handshakes between the remaining $n - 1$ persons (which is $(n-1)(n-2)/2$ by the induction hypothesis). We get $(n-1) + (n-1)(n-2)/2 = n(n-1)/2$ handshakes.

2.1.12. We did not check the base case $n = 1$.

2.1.13. The proof uses that there are at least four lines. But we checked only $n = 1, 2$ as base cases. The assertion is false for $n = 3$ and for every value of n after that.

2.2 Comparing and Estimating Numbers

2.2.1. (a) The left-hand side counts all subsets of an n-set; the right-hand side counts only the 3-element subsets. (b) $2^n/n^2 > \binom{n}{3}/n^2 = (n-1)(n-2)/(6n)$, which becomes arbitrarily large.

2.2.2. Start the induction with $n = 4$: $4! = 24 > 16 = 2^4$. If the inequality holds for n, then $(n+1)! = (n+1)n! > (n+1)2^n > 2 \cdot 2^n = 2^{n+1}$.

2.3 Inclusion–Exclusion

2.3.1. $18 + 23 + 21 + 17 - 9 - 7 - 6 - 12 - 9 - 12 + 4 + 3 + 5 + 7 - 3 = 40$.

2.4 Pigeonholes

2.4.1. If each of the giant boxes contains at most 20 New Yorkers, then 500,000 boxes contain at most $20 \cdot 500,000 = 10,000,000$ New Yorkers, which is a contradiction.

3 Binomial Coefficients and Pascal's Triangle

3.1 The Binomial Theorem

3.1.1.

$$(x + y)^{n+1}$$
$$= (x + y)^n (x + y)$$
$$= \left(x^n + \binom{n}{1} x^{n-1} y + \cdots + \binom{n}{n-1} xy^n + \binom{n}{n} y^n \right) (x + y)$$
$$= x^n (x + y) + \binom{n}{1} x^{n-1} y (x + y) + \cdots$$
$$\quad + \binom{n}{n-1} xy^{n-1} (x + y) + \binom{n}{n} y^n (x + y)$$
$$= \left(x^{n+1} + x^n y \right) + \binom{n}{1} \left(x^n y + x^{n-1} y^2 \right) + \cdots$$
$$\quad + \binom{n}{n-1} \left(x^2 y^{n-1} + xy^n \right) + \binom{n}{n} \left(xy^n + y^{n+1} \right)$$
$$= x^{n+1} + \left(1 + \binom{n}{1} \right) x^n y + \left(\binom{n}{1} + \binom{n}{2} \right) x^{n-1} y^2 + \cdots$$
$$\quad + \left(\binom{n}{n-1} + \binom{n}{n} \right) xy^n + y^{n+1}$$
$$= x^{n+1} + \binom{n+1}{1} x^n y + \binom{n+1}{2} x^{n-1} y^2 + \cdots + \binom{n+1}{n} xy^n + y^{n+1}.$$

3.1.2. (a) $(1 - 1)^n = 0$. (b) By $\binom{n}{k} = \binom{n}{n-k}$.

3.1.3. The identity says that *the number of subsets of an n-element set with an even number of elements is the same as the number of subsets with an odd number of elements*. We can establish a bijection between even and odd subsets as follows: If a subset contains 1, delete it from the subset; otherwise, add it to the subset.

3.2 Distributing Presents

3.2.1.

$$\binom{n}{n_1} \cdot \binom{n - n_1}{n_2} \cdots \binom{n - n_1 - \cdots - n_{k-1}}{n_k}$$
$$= \frac{n!}{n_1!(n - n_1)!} \frac{(n - n_1)!}{n_2!(n - n_1 - n_2)!} \cdots \frac{(n - n_1 - \cdots - n_{k-1})!}{n_{k-1}!(n - n_1 - \cdots - n_k)!}$$
$$= \frac{n!}{n_1! n_2! \cdots n_k!},$$

since $n - n_1 - \cdots - n_{k-1} - n_k = 0$.

3.2.2. (a) $n!$ (distribute positions instead of presents). (b) $n(n-1)\cdots(n-k+1)$ (distribute as "presents" the first k positions at the competition and $n - k$ certificates of participation). (c) $\binom{n}{n_1}$. (d) Chess seating in Diane's sense (distribute players to boards).

3.2.3. (a) $[n = 8]$ $8!$. (b) $8! \cdot \binom{8}{4}$. (c) $(8!)^2$.

3.3 Anagrams

3.3.1. $13!/2^3$.

3.3.2. COMBINATORICS.

3.3.3. Most: any word with 13 different letters; least: any word with 13 identical letters.

3.3.4. (a) 26^6.

(b) $\binom{26}{4}$ ways to select the four letters that occur; for each selection, $\binom{4}{2}$ ways to select the two letters that occur twice; for each selection, we distribute 6 positions to these letters (2 of them get 2 positions); this gives $\frac{6!}{2!2!}$ ways. Thus we get $\binom{26}{4}\binom{4}{2}\frac{6!}{2!2!}$. (There are many other ways to arrive at the same number!)

(c) Number of ways to partition 6 into a sum of positive integers:

$$6 = 6 = 5 + 1 = 4 + 2 = 4 + 1 + 1 = 3 + 3 = 3 + 2 + 1 = 3 + 1 + 1 + 1$$

$$= 2 + 2 + 2 = 2 + 2 + 1 + 1 = 2 + 1 + 1 + 1 + 1 = 1 + 1 + 1 + 1 + 1 + 1,$$

which makes 11 possibilities.

(d) This is too difficult in this form. What we meant is the following: how many words of length n are there such that none is an anagram of another? This means distributing n pennies to 26 children, and so the answer is $\binom{n+25}{25}$.

3.4 Distributing Money

3.4.1. $\binom{n-k-1}{k-1}$.

3.4.2. $\binom{n+k-1}{\ell+k-1}$.

3.4.3. $\binom{kp+k-1}{k-1}$.

3.5 Pascal's Triangle

3.5.1. This is the same as $\binom{n}{k} = \binom{n}{n-k}$.

3.5.2. $\binom{n}{0} = \binom{n}{n} = 1$ (e.g., by the general formula for the binomial coefficients).

3.6 Identities in Pascal's Triangle

3.6.1.

$$1 + \binom{n}{1} + \binom{n}{2} + \cdots + \binom{n}{n-1} + \binom{n}{n}$$

$$= 1 + \left[\binom{n-1}{0} + \binom{n-1}{1}\right] + \left[\binom{n-1}{1} + \binom{n-1}{2}\right] +$$

$$\cdots + \left[\binom{n-1}{n-2} + \binom{n-1}{n-1}\right] + 1$$

$$= 2\left[\binom{n-1}{0} + \binom{n-1}{1} + \cdots + \binom{n-1}{n-2} + \binom{n-1}{n-1}\right]$$

$$= 2 \cdot 2^{n-1} = 2^n.$$

3.6.2. The coefficient of $x^n y^n$ in

$$\left(\binom{n}{0}x^n + \binom{n}{1}x^{n-1}y + \cdots + \binom{n}{n-1}xy^{n-1} + \binom{n}{n}y^n\right)^2$$

is

$$\binom{n}{0}\binom{n}{n} + \binom{n}{1}\binom{n}{n-1} + \cdots + \binom{n}{n-1}\binom{n}{1} + \binom{n}{n}\binom{n}{0}.$$

3.6.3. The left-hand side counts all k-element subsets of an $(n+m)$-element set by distinguishing them according to how many elements they pick up from the first n.

3.6.4. If the largest element is j (which is at least $n+1$), then the rest can be chosen $\binom{j-1}{n}$ ways. If we sum over all $j \geq n+1$, we get the identity

$$\binom{n}{n} + \binom{n+1}{n} + \cdots + \binom{n+k}{n} = \binom{n+k+1}{n+1}.$$

Using that $\binom{n+i}{n} = \binom{n+i}{i}$, we get (3.5).

3.7 A Bird's Eye View of Pascal's Triangle

3.7.1. $n = 3k + 2$.

3.7.2. This is not easy. We want to determine the first value of k where the difference of differences turns nonpositive:

$$\left(\binom{n}{k+1} - \binom{n}{k}\right) - \left(\binom{n}{k} - \binom{n}{k-1}\right) \leq 0.$$

We can divide the equation by $\binom{n}{k-1}$ and multiply by $k(k+1)$ to get

$$(n - k + 1)(n - k) - 2(n - k + 1)(k + 1) + k(k + 1) \leq 0.$$

Simplifying, we obtain

$$4k^2 - 4nk + n^2 - n - 2 < 0.$$

Solving for k, we get that the left-hand side is nonpositive between the two roots:

$$\frac{n}{2} - \frac{1}{2}\sqrt{n + 2} \leq k \leq \frac{n}{2} + \frac{1}{2}\sqrt{n + 2}.$$

So the first integer k for which this is nonpositive is

$$k = \left\lceil \frac{n}{2} - \frac{1}{2}\sqrt{n + 2} \right\rceil.$$

3.8 An Eagle's-Eye View: Fine Details

3.8.2. We apply the lower bound in Lemma 2.5.1 to the logarithms. For a typical term, we get

$$\ln\left(\frac{m + t - k}{m - k}\right) \geq \frac{\frac{m+t-k}{m-k} - 1}{\frac{m+t-k}{m-k}} = \frac{t}{m + t - k},$$

and so

$$\ln\left(\frac{m + t}{m}\right) + \ln\left(\frac{m + t - 1}{m - 1}\right) + \cdots + \ln\left(\frac{m + 1}{m - t + 1}\right)$$

$$\geq \frac{t}{m + t} + \frac{t}{m + t - 1} + \cdots + \frac{t}{m + 1}.$$

We replace each denominator by the *largest* one to decrease the sum:

$$\frac{t}{m + t} + \frac{t}{m + t - 1} + \cdots + \frac{t}{m + 1} \geq \frac{t^2}{m + t}.$$

Inverting the steps of taking the logarithm and taking the reciprocal, this gives the upper bound in (3.9).

3.8.1. (a) We have to show that $e^{-t^2/(m-t+1)} \leq e^{-t^2/m} \leq e^{-t^2/(m+t)}$. This is straightforward using that e^x is a monotone increasing function.

(b) Take the ratio of the upper and lower bounds; we obtain

$$\frac{e^{-t^2/(m+t)}}{e^{-t^2/(m-t+1)}} = e^{t^2/(m-t+1) - t^2/(m+t)}.$$

Here the exponent is

$$\frac{t^2}{m-t+1} - \frac{t^2}{m+t} = \frac{(2t-1)t^2}{(m-t+1)(m+t)}.$$

In our case, this is $1900/(41*60) \approx 0.772$, and so the ratio is $e^{0.772} \approx 2.1468$.

3.8.3. By (3.9), we have

$$\binom{2m}{m} \Big/ \binom{2m}{m-t} \geq e^{t^2/(m+t)}.$$

Here the exponent is a monotone increasing function of t for $t \geq 0$ (to see this, write it as $t(1 - \frac{m}{m+t})$, or take its derivative), and so from our assumption that $t \geq \sqrt{m \ln C} + \ln C$ it follows that

$$\frac{t^2}{m+t} \geq \frac{(\sqrt{m \ln C} + \ln C)^2}{m + \sqrt{m \ln C} + \ln C} = \frac{\ln C(m + 2\sqrt{m \ln C} + \ln C)}{m + \sqrt{m \ln C} + \ln C}$$
$$> \ln C,$$

which implies that

$$\binom{2m}{m} \Big/ \binom{2m}{m-t} > C.$$

The proof of the other half is similar.

4 Fibonacci Numbers

4.1 Fibonacci's Exercise

4.1.1. Because we use the two previous elements to compute the next.

4.1.2. F_{n+1}.

4.1.3. Let us denote by S_n the number of good subsets. If $n = 1$, then $S_1 = 2$ (the empty set and the set $\{1\}$. If $n = 2$, then $\emptyset$, $\{1\}$, $\{2\}$, so $S_2=3$. For any n, if the subset contains n, then it can not contain $n - 1$, so there are S_{n-2} subsets of this type; if it does not contain n, then there are S_{n-1} subsets. So we have the same recursive formula, so $S_n = F_{n+2}$.

4.2 Lots of Identities

4.2.1. It is clear from the recurrence that two odd members are followed by an even, then by two odd numbers again.

4.2.2. We formulate the following nasty-looking statement: *If n is divisible by 5, then so is F_n; if n has remainder 1 when divided by 5, then F_n has remainder 1; if n has remainder 2 when divided by 5, then F_n has remainder*

1; *if n has remainder* 3 *when divided by* 5, *then* F_n *has remainder* 2; *if n has remainder* 4 *when divided by* 5, *then* F_n *has remainder* 3. This is then easily proved by induction on n.

4.2.3. By induction. All of them are true for $n = 1$ and $n = 2$. Assume that $n \geq 3$.

(a) $F_1 + F_3 + F_5 + \cdots + F_{2n-1} = (F_1 + F_3 + \cdots + F_{2n-3}) + F_{2n-1} = F_{2n-2} + F_{2n-1} = F_{2n}$.

(b) $F_0 - F_1 + F_2 - F_3 + \cdots - F_{2n-1} + F_{2n} = (F_0 - F_1 + F_2 - \cdots + F_{2n-2}) + (-F_{2n-1} + F_{2n}) = (F_{2n-3} - 1) + F_{2n-2} = F_{2n-1} - 1$.

(c) $F_0^2 + F_1^2 + F_2^2 + \cdots + F_n^2 = (F_0^2 + F_1^2 + \cdots + F_{n-1}^2) + F_n^2 = F_{n-1}F_n + F_n^2 = F_n(F_{n-1} + F_n) = F_n \cdot F_{n+1}$.

(d) $F_{n-1}F_{n+1} - F_n^2 = F_{n-1}(F_{n-1} + F_n) - F_n^2 = F_{n-1}^2 + F_n(F_{n-1} - F_n) = F_{n-1}^2 - F_nF_{n-2} = -(-1)^{n-1} = (-1)^n$.

4.2.4. We can write (4.1) as $F_{n-1} = F_{n+1} - F_n$, and use this to compute F_n for negative n recursively (going backwards):

$$\ldots, -21, 13, -8, 5, -3, 2, -1, 1, 0.$$

It is easy to recognize that these are the same as the ordinary Fibonacci numbers, except that every second number has a negative sign. As a formula, we have

$$F_{-n} = (-1)^{n+1} F_n.$$

This is now easily proved by induction on n. It is true for $n = 0, 1$, and assuming that it is true for n and $n - 1$, we get for $n + 1$,

$$F_{-(n+1)} = F_{-(n-1)} - F_{-n} = (-1)^n F_{n-1} - (-1)^{n+1} F_n$$
$$= (-1)^n (F_{n-1} + F_n) = (-1)^n F_{n+1} = (-1)^{n+2} F_{n+1},$$

which completes the induction.

4.2.5.

$$F_{n+2} = F_{n+1} + F_n = (F_n + F_{n-1}) + F_n = 2F_n + (F_n - F_{n-2}) = 3F_n - F_{n-2}.$$

Replacing n by $2n - 1$, we get the recurrence for odd-index Fibonacci numbers. Using this to prove (4.2):

$$F_{n+1}^2 + F_n^2 = (F_n + F_{n-1})^2 + F_n^2 = 2F_n^2 + F_{n-1}^2 + 2F_nF_{n-1}$$
$$= 3F_n^2 + 2F_{n-1}^2 - (F_n - F_{n-1})^2 = 3F_n^2 + 2F_{n-1}^2 - F_{n-2}^2$$
$$= 3(F_n^2 + F_{n-1}^2) - (F_{n-1}^2 + F_{n-2}^2) = 3F_{2n-1} - F_{2n-3}$$
$$= F_{2n+1}.$$

4.2.6. The identity is

$$\binom{n}{0} + \binom{n-1}{1} + \binom{n-2}{2} + \cdots + \binom{n-k}{k} = F_{n+1},$$

where $k = \lfloor n/2 \rfloor$. Proof by induction. True for $n = 0$ and $n = 1$. Let $n \geq 2$. Assume that n is odd; the even case is similar, except that the last term below needs a little different treatment.

$$\binom{n}{0} + \binom{n-1}{1} + \binom{n-2}{2} + \cdots + \binom{n-k}{k}$$

$$= 1 + \left(\binom{n-2}{0} + \binom{n-2}{1} \right) + \left(\binom{n-3}{1} + \binom{n-3}{2} \right) + \cdots$$

$$+ \left(\binom{n-k-1}{k-1} + \binom{n-k-1}{k} \right)$$

$$= \left(\binom{n-1}{0} + \binom{n-2}{1} + \binom{n-3}{2} + \cdots + \binom{n-k-1}{k} \right)$$

$$+ \left(\binom{n-2}{0} + \binom{n-3}{1} + \cdots + \binom{n-k-1}{k-1} \right)$$

$$= F_n + F_{n-1} = F_{n+1}.$$

4.2.7. (4.2) follows by taking $a = b = n - 1$. (4.3), follows by taking $a = n$, $b = n - 1$.

4.2.8. Let $n = km$. We use induction on m. For $m = 1$ the assertion is obvious. If $m > 1$, then we use (4.5) with $a = k(m-1)$, $b = k - 1$:

$$F_{ka} = F_{(k-1)a}F_{a-1} + F_{(k-1)a+1}F_a.$$

By the induction hypothesis, both terms are divisible by F_a.

4.2.9. The "diagonal" is in fact a very long and narrow parallelogram with area 1. The trick depends on the fact $F_{n+1}F_{n-1} - F_n^2 = (-1)^n$ is very small compared to F_n^2.

4.3 A Formula for the Fibonacci Numbers

4.3.1. True for $n = 0, 1$. Let $n \geq 2$. Then by the induction hypothesis,

$$F_n = F_{n-1} + F_{n-2}$$

$$= \frac{1}{\sqrt{5}} \left(\left(\frac{1+\sqrt{5}}{2} \right)^{n-1} - \left(\frac{1-\sqrt{5}}{2} \right)^{n-1} \right)$$

$$+ \frac{1}{\sqrt{5}} \left(\left(\frac{1+\sqrt{5}}{2} \right)^{n-2} - \left(\frac{1-\sqrt{5}}{2} \right)^{n-2} \right)$$

$$= \frac{1}{\sqrt{5}} \left(\left(\frac{1+\sqrt{5}}{2} \right)^{n-2} \left(\frac{1+\sqrt{5}}{2} + 1 \right) + \left(\frac{1-\sqrt{5}}{2} \right)^{n-2} \left(\frac{1-\sqrt{5}}{2} + 1 \right) \right)$$

$$= \frac{1}{\sqrt{5}} \left(\left(\frac{1+\sqrt{5}}{2} \right)^{n} - \left(\frac{1-\sqrt{5}}{2} \right)^{n} \right).$$

4.3.2. For $n = 1$ and $n = 2$, if we require that L_n be of the given form, then we get

$$L_1 = 1 = a + b, \qquad L_2 = 3 = a \frac{1+\sqrt{5}}{2} + b \frac{1-\sqrt{5}}{2}.$$

Solving for a and b, we get

$$a = \frac{1+\sqrt{5}}{2}, \qquad b = \frac{1-\sqrt{5}}{2}.$$

Then

$$L_n = \left(\frac{1+\sqrt{5}}{2} \right)^{n} + \left(\frac{1-\sqrt{5}}{2} \right)^{n},$$

which follows by induction on n just as in the previous problem.

4.3.3. (a) For example, every day Jack buys either an ice cream for \$1 or a giant sundae for \$2. There are 4 different flavors of ice cream, but only one kind of sundae. If he has n dollars, in how many ways can he spend the money?

$$I_n = \frac{1}{2\sqrt{5}} \left((2 + \sqrt{5})^n - (2 - \sqrt{5})^n \right).$$

4.3.4. The formula works for $n = 1, 2, \ldots, 10$ but fails for $n = 11$, when it gives 91. In fact, it will be more and more off as we increase n. We have seen that

$$F_n \sim \frac{1}{\sqrt{5}} \left(\frac{1+\sqrt{5}}{2} \right)^{n} = (0.447 \ldots) \cdot (1.618 \ldots)^n.$$

In the formula of Alice, the rounding plays less and less of a role, so

$$\left\lceil e^{n/2-1} \right\rceil \sim e^{n/2-1} = (0.367\dots) \cdot (1.648\dots)^n,$$

and so the ratio between Alice's numbers and the corresponding Fibonacci numbers is

$$\frac{\left\lceil e^{n/2-1} \right\rceil}{F_n} \approx \frac{(0.367\dots) \cdot (1.648\dots)^n}{0.447\dots} = (0.822\dots) \cdot (1.018\dots)^n.$$

Since the base of the exponential is larger than 1, this tends to infinity as n grows.

5 Combinatorial Probability

5.1 Events and Probabilities

5.1.1. The union of two events A and B corresponds to "A or B", i.e., at least one of A or B occurs.

5.1.2. It is the sum of some of the probabilities of outcomes, and even if we add up all of the probabilities, we get just 1.

5.1.3. $P(E) = \frac{1}{2}$, $P(T) = \frac{1}{3}$.

5.1.4. The same probabilities $P(s)$ are added up on both sides.

5.1.5. Every probability $P(s)$ with $s \in A \cap B$ is added twice to both sides; every probability $P(s)$ with $s \in A \cup B$ but $s \notin A \cap B$ is added once to both sides.

5.2 Independent Repetition of an Experiment

5.2.1. The pairs $(E,T), (O,T), (L,T)$ are independent. The pair (E,O) is exclusive. Neither the pair (E,L) nor the pair (O,L) is independent.

5.2.2. $P(\emptyset \cap A) = P(\emptyset) = 0 = P(\emptyset)P(A)$. The set S also has this property: $P(S \cap A) = P(A) = P(S)P(A)$.

5.2.3. $P(A) = \frac{|S|^{n-1}}{|S|^n} = \frac{1}{|S|}$, $P(B) = \frac{|S|^{n-1}}{|S|^n} = \frac{1}{|S|}$, $P(A \cap B) = \frac{|S|^{n-2}}{|S|^n} = \frac{1}{|S|^2} = P(A)P(B)$.

5.2.4. The probability that your mother has the same birthday as you is $1/365$ (here we assume that birthdays are distributed evenly among all numbers of the year, and we ignore leap years). There are (roughly) 7 billion people in the word. This means that one could expect that $7 \cdot 10^9/365$ (about 20 million) people have the same birthday as their mother. The events that your birthday coincides with that of your mother, father, or spouse are independent, so the probability that for a given person, all three were born on his or her birthday is $1/365^3 = 1/48{,}627{,}125$. Let's say there are 2 billion married people; then we can expect that $2{,}000{,}000{,}000/48{,}627{,}125 \approx 41$ of them have the same birthday as their mother, father, and spouse.

6 Integers, Divisors, and Primes

6.1 Divisibility of Integers

6.1.1. $a = a \cdot 1 = (-a) \cdot (-1)$.

6.1.2. (a) even; (b) odd; (c) $a = 0$.

6.1.3. (a) If $b = am$ and $c = bn$, then $c = amn$. (b) If $b = am$ and $c = an$, then $b + c = a(m + n)$ and $b - c = a(m - n)$. (c) If $b = am$ and $a, b > 0$, then $m > 0$; hence $m \geq 1$, and so $b \geq a$. (d) Trivial if $a = 0$. Assume $a \neq 0$. If $b = am$ and $a = bn$, then $a = amn$, so $mn = 1$. Hence either $m = n = 1$ or $m = n = -1$.

6.1.4. We have $a = cn$ and $b = cm$, hence $r = b - aq = c(m - nq)$.

6.1.5. We have $b = am$, $c = aq + r$ and $c = bt + s$. Hence $s = c - bt = (aq + r) - (am)t = (q - mt)a + r$. Since $0 \leq r < a$, the remainder of the division $s \div a$ is r.

6.1.6. (a) $a^2 - 1 = (a - 1)(a + 1)$. (b) $a^n - 1 = (a - 1)(a^{n-1} + \cdots + a + 1)$.

6.3 Factorization into Primes

6.3.1. There is a smallest one among *positive* criminals (indeed, in every set of positive integers), but a set of negative integers need not have a smallest element (if it is infinite).

6.3.2. Yes, the number 2.

6.3.3. (a) p occurs in the prime factorization of ab, so it must occur in the prime factorization of a or in the prime factorization of b.

(b) $p \mid a(b/a)$, but $p \nmid a$, so by (a), we must have $p \mid (b/a)$.

6.3.4. Let $n = p_1 p_2 \cdots p_k$; each $p_i \geq 2$; hence $n \geq 2^k$.

6.3.5. If $r_i = r_j$, then $ia - ja$ is divisible by p. But $ia - ja = (i - j)a$ and neither a nor $i - j$ is divisible by p. Hence the r_i are all different. None of them is 0. Their number is $p - 1$, so every value $1, 2, \ldots, p - 1$ must occur among the r_i.

6.3.6. For a prime p, the proof is the same as for 2. If n is composite but not a square, then there is a prime p that occurs in the prime factorization of n an odd number of times. We can repeat the proof by looking at this p.

6.3.7. Fact: If $\sqrt[k]{n}$ is not an integer, then it is irrational. Proof: There is a prime that occurs in the prime factorization of n, say t times, where $k \nmid t$. If (indirect assumption) $\sqrt[k]{n} = a/b$ then $nb^k = a^k$, and so the number of times p occurs in the prime factorization of the left-hand side is not divisible by k, while the number of times it occurs in the prime factorization of the right-hand side is divisible by k. A contradiction.

6.4 On the Set of Primes

6.4.1. Just as in the treatment of the case $k = 200$ above, we subtract the number of primes up to 10^{k-1} from the number of primes up to 10^k. By the Prime Number Theorem, this number is about

$$\frac{10^k}{k \ln 10} - \frac{10^{k-1}}{(k-1) \ln 10} = \frac{(9k-10)10^{k-1}}{k(k-1) \ln 10}.$$

Since

$$\frac{9k-10}{k-1} = 9 - \frac{1}{k-1}$$

is very close to 9 if k is large, we get that the number of primes with k digits is approximately

$$\frac{9 \cdot 10^{k-1}}{k \ln 10}.$$

Comparing this with the total number of positive integers with k digits, which we know is $10^k - 10^{k-1} = 9 \cdot 10^{k-1}$, we get

$$\frac{9 \cdot 10^{k-1}}{k \ln 10 \cdot 9 \cdot 10^{k-1}} = \frac{1}{(\ln 10)k} \approx \frac{1}{2.3k}.$$

6.5 Fermat's "Little" Theorem

6.5.1. $4 \nmid \binom{4}{2} = 6.$ $4 \nmid 2^4 - 2 = 14.$

6.5.2. (a) We need that each of the p rotated copies of a set are different. Suppose that there is a rotated copy that occurs a times. Then trivially every other rotated copy occurs a times. But then $a \mid p$, so we must have $a = 1$ or $a = p$. If all p rotated copies are the same, then trivially either $k = 0$ or $k = p$, which were excluded. So we have $a = 1$ as claimed. (b) Consider the set of two opposite vertices of a square. (c) If each box contains p subsets of size k, the total number of subsets must be divisible by p.

6.5.3. We consider each number to have p digits by appending zeros at the front if necessary. We get p numbers from each number a by a cyclic shift. These are all the same when all digits of a are the same, but all different otherwise (why? the assumption that p is a prime is needed here!). So we get $a^p - a$ numbers that are divided into classes of size p. Thus $p \mid a^p - a$.

6.5.4. Assume that $p \nmid a$. Consider the product $a(2a)(3a) \cdots ((p-1)a) = (p-1)! a^{p-1}$. Let r_i be the remainder of ia when divided by p. Then the product above has the same remainder when divided by p as the product $r_1 r_2 \cdots r_{p-1}$. But this product is just $(p-1)!$. Hence p is a divisor of $(p-1)! a^{p-1} - (p-1)! = (p-1)!(a^{p-1} - 1)$. Since p is a prime, it is not a divisor of $(p-1)!$, and so it is a divisor of $a^{p-1} - 1$.

6.6 The Euclidean Algorithm

6.6.1. $\gcd(a, b) \le a$, but a is a common divisor, so $\gcd(a, b) = a$.

6.6.2. (a) Let $d = \gcd(a, b)$. Then $d \mid a$ and $d \mid b$, and hence $d \mid b - a$. Thus d is a common divisor of a and $b - a$, and hence $d \le \gcd(a, b)$. A similar argument shows the reverse inequality. (b) By repeated application of (a).

6.6.3. (a) $\gcd(a/2, b) \mid (a/2)$ and hence $\gcd(a/2, b) \mid a$. So $\gcd(a/2, b)$ is a common divisor of a and b, and hence $\gcd(a/2, b) \le \gcd(a, b)$. The reverse inequality follows similarly, using that $\gcd(a, b)$ is odd, and hence $\gcd(a, b) \mid (a/2)$.

(b) $\gcd(a/2, b/2) \mid (a/2)$ and hence $2\gcd(a/2, b/2) \mid a$. Similarly, we have $2\gcd(a/2, b/2) \mid b$, and hence $2\gcd(a/2, b/2) \le \gcd(a, b)$. Conversely, $\gcd(a, b) \mid a$ and hence $\frac{1}{2}\gcd(a, b) \mid a/2$. Similarly, $\frac{1}{2}\gcd(a, b) \mid b/2$, and hence $\frac{1}{2}\gcd(a, b) \le \gcd(a/2, b/2)$.

6.6.4. Consider each prime that occurs in either one of them, raise it to the larger of the two exponents, and multiply these prime powers.

6.6.5. If a and b are the two integers, and you know the prime factorization of a, then take a prime factor of a, divide b by this prime repeatedly to determine its exponent in the prime factorization of b, and raise this prime to the smaller of its exponent in the prime factorizations of a and b. Repeat this for all prime factors of a, and multiply these prime powers.

6.6.6. By the descriptions of the gcd and lcm above, each prime occurs the same number of times in the prime factorization of both sides.

6.6.7. (a) Straightforward. (b) Let $z = \gcd(a, b, c)$, and let $A = a/z$, $B = b/z$, $C = c/z$. Then A, B, and C are relatively prime and form a Pythagorean triple. One of A and B must be odd, since if both of them were even, then C would be even as well, and so the three numbers would not be relatively prime. Suppose that B is odd. Then A must be even. Indeed, the square of an odd number gives a remainder of 1 when divided by 4, so if both A and B were odd, then $C^2 = A^2 + B^2$ would give a remainder of 2 when divided by 4, which is impossible. It follows that C must be odd.

So A is even, and we can write it in the form $A = 2A_0$. Write the equation in the form

$$A_0^2 = \frac{C+B}{2} \frac{C-B}{2}.$$

Let p be any prime number dividing A_0. Then p must divide either $(C + B)/2$ or $(C - B)/2$. But p cannot divide both, since then it would also divide the sum $\frac{C+B}{2} + \frac{C-B}{2} = C$ as well as the difference $\frac{C+B}{2} - \frac{C-B}{2} = B$, contradicting the assumption that A, B, and C are relatively prime.

The prime p may occur in the prime decomposition of A_0 several times, say k times. Then in the prime decomposition of A_0^2, p occurs $2k$ times. By the argument above, p must occur $2k$ times in the prime decomposition of one of $(C+B)/2$ and $(C-B)/2$, and not at all in the prime decomposition of the other. So we see that in the prime decomposition of $(C+B)/2$ (and similarly in the prime decomposition of $(C - B)/2$), every prime occurs to an even power. This is the same as saying that both $(C + B)/2$ and $(C - B)/2$ are squares; say, $(C + B)/2 = x^2$ and $(C - B)/2 = y^2$ for some integers x and y.

Now we can express A, B, and C in terms of x and y:

$$B = \frac{C + B}{2} - \frac{C - B}{2} = x^2 - y^2, \quad C = \frac{C + B}{2} + \frac{C - B}{2} = x^2 + y^2,$$

$$A = 2A_0 = 2\sqrt{\frac{C + B}{2}\frac{C - B}{2}} = 2xy.$$

We get a, b, and c by multiplying A, B, and C by z, which completes the solution.

6.6.8. $\gcd(a, a + 1) = \gcd(a, 1) = \gcd(0, 1) = 1$.

6.6.9. The remainder of F_{n+1} divided by F_n is F_{n-1}. Hence $\gcd(F_{n+1}, F_n) = \gcd(F_n, F_{n-1}) = \cdots = \gcd(F_3, F_2) = 1$. This lasts $n - 1$ steps.

6.6.10. By induction on k. True if $k = 1$. Suppose that $k > 1$. Let $b = aq + r$, $1 \le r < a$. Then the Euclidean Algorithm for computing $\gcd(a, r)$ lasts $k - 1$ steps; hence $a \ge F_k$ and $r \ge F_{k-1}$ by the induction hypothesis. But then $b = aq + r \ge a + r \ge F_k + F_{k-1} = F_{k+1}$.

6.6.11. (a) Takes 10 steps. (b) Follows from $\gcd(a, b) = \gcd(a - b, b)$. (c) $\gcd(10^{100} - 1, 10^{100} - 2)$ takes $10^{100} - 1$ steps.

6.6.12. (a) Takes 8 steps. (b) At least one of the numbers remains odd all the time. (c) Follows from Exercises 6.6.2 and 6.6.3. (d) The product of the two numbers drops by a factor of two in one of any two iterations.

6.7 Congruences

6.7.1. $m = 54321 - 12345 = 41976$.

6.7.2. Only (b) is correct.

6.7.3. $a \equiv b \pmod{0}$ should mean that there exists an integer k such that $a - b = 0 \cdot k$. This means that $a - b = 0$, that is, $a = b$. So equality can be considered as a special case of congruence.

6.7.4. (a) Take $a = 2$ and $b = 5$. (b) If $ac \equiv bc \pmod{mc}$, then $mc \mid ac - bc$, so there is an integer k such that $ac - bc = kmc$. Since $c \neq 0$, this implies that $a - b = km$, and so $a \equiv b \pmod{m}$.

6.7.5. First, from $x \equiv y \pmod{p}$ it follows (by the multiplication rule) that $x^v \equiv y^v \pmod{p}$, so it suffices to prove that

$$x^u \equiv x^v \pmod{p}. \tag{16.1}$$

If $x \equiv 0 \pmod{p}$, then both sides of (16.1) are divisible by p, and the assertion follows. Suppose that $x \not\equiv 0 \pmod{p}$. Suppose that (say) $u < v$. We know that $p - 1 \mid v - u$, so we can write $v - u = k(p - 1)$ with some positive integer k. Now we know by Fermat's Little Theorem that $x^{p-1} \equiv 1 \pmod{p}$, hence by the multiplication rule of congruences, we have $x^{k(p-1)} \equiv 1 \pmod{p}$, and by the multiplication rule again, we get $x^v = x^u \cdot x^{k(p-1)} \equiv x^u \pmod{p}$, which proves (16.1).

6.8 Strange Numbers

6.8.1. Tu; Sa; Th; We.

6.8.2. not-$A = 1 \oplus A$; A-or-$B = A \oplus B \oplus A \cdot B$; A-and-$B = A \cdot B$.

6.8.3. $2 \cdot 0 \equiv 2 \cdot 3 \pmod{6}$ but $0 \not\equiv 3 \pmod{6}$. More generally, if $m = ab$ $(a, b > 1)$ is a composite modulus, then $a \cdot 0 \equiv a \cdot b \pmod{m}$, but $0 \not\equiv b \pmod{m}$.

6.8.4. We start with the Euclidean Algorithm:

$$\gcd(53, 234527) = \gcd(53, 2) = \gcd(1, 2) = 1.$$

Here we got 2 as $2 = 234527 - 4425 \cdot 53$, and then 1 as

$$1 = 53 - 26 \cdot 2 = 53 - 26(234527 - 4425 \cdot 53) = 115051 \cdot 53 - 26 \cdot 234527.$$

It follows that $1 \equiv 115051 \cdot 53 \pmod{234527}$, and so $\overline{1/53} = \overline{115051}$.

6.8.5. $x \equiv 5$, $y \equiv 8 \pmod{11}$.

6.8.6. (a) We have $11 \mid x^2 - 2x = x(x - 2)$; hence either $11 \mid x$ or $11 \mid x - 2$, so $x \equiv 0 \pmod{11}$ and $x \equiv 2 \pmod{11}$ are the two solutions. (b) Similarly from $23 \mid x^2 - 4 = (x - 2)(x + 2)$ we get $x \equiv 2 \pmod{23}$ or $x \equiv -2 \pmod{23}$.

6.9 Number Theory and Combinatorics

6.9.1. There are two neighboring integers k and $k + 1$ among the given n numbers (Pigeonhole Principle) that are relatively prime.

6.9.2. By the rules of inclusion-exclusion, we have to subtract from n the number of multiples of p_i (between 1 and n) for every p_i; then we have to add the number of common multiples of p_i and p_j for any two primes p_i and p_j; then we have to subtract the number of common multiples of p_i, p_j, and p_k for any three primes p_i, p_j, and p_k, etc. Just as in the numerical example, the number of multiples of p_i is n/p_i; the number of common multiples of p_i and p_j is $n/(p_i p_j)$; the number of common multiples of p_i, p_j, and p_k is $n/(p_i p_j p_k)$, etc. So we get

$$\phi(n) = n - \frac{n}{p_1} - \cdots - \frac{n}{p_r} + \frac{n}{p_1 p_2} + \frac{n}{p_1 p_3} + \cdots + \frac{n}{p_{r-1} p_r} - \frac{n}{p_1 p_2 p_2} - \cdots .$$

This is equal to the expression in (6.7). Indeed, if we expand the product, every term arises by picking either "1" or "$-\frac{1}{p_i}$" from each factor $\left(1 - \frac{1}{p_i}\right)$, which gives a term of the form

$$(-1)^k \frac{1}{p_{i_1} \cdots p_{i_k}} .$$

This is just a typical term in the inclusion-exclusion formula above.

6.9.3. It is not hard to come up with the conjecture that the answer is n. To prove this, consider the fractions $\frac{1}{n}, \frac{2}{n}, \ldots, \frac{n}{n}$, and simplify them as much as possible. We get fractions of the form $\frac{a}{d}$, where d is a divisor of n, $1 \le a \le d$, and $\gcd(a, d) = 1$. It is also clear that we get every such fraction. The number of such fractions with a given denominator is $\phi(d)$. Since the total number of fractions we started with is n, this proves our conjecture.

6.9.4. For $n = 1$ and 2 the answer is 1. Suppose that $n > 2$. If k is such an integer, then so is $n - k$. So these integers come in pairs adding up to n (we have to add that $n/2$ is not among these numbers). There are $\phi(n)/2$ such pairs, so the answer is $n\phi(n)/2$.

6.9.5. The proof is similar to the solution of Exercise 6.5.4.

Let $s_1, \ldots, s_k$ be the numbers between 1 and b relatively prime to b; so $k = \phi(b)$. Let r_i be the remainder of $s_i a$ when divided by p. We have $\gcd(b, r_i) = 1$, since if there were a prime p dividing both b and r_i, then p would also divide $s_i a$, which is impossible, since both s_i and a are relatively prime to b. Second, $r_1, r_2, \ldots, r_k$ are different, since $r_i = r_j$ would mean that $b \mid s_i a - s_j a = (s_i - s_j)a$; since $\gcd(a, b) = 1$, this would imply that $b \mid s_i - s_j$, which is clearly impossible. Hence it follows that $r_1, r_2, \ldots, r_k$ are just the numbers $s_1, s_2, \ldots, s_k$, in a different order.

Consider the product $(s_1 a)(s_2 a) \cdots (s_k a)$. On the one hand, we can write this as

$$(s_1 a)(s_2 a) \cdots (s_k a) = (s_1 s_2 \cdots s_k)a^k ,$$

on the other,

$$(s_1 a)(s_2 a) \cdots (s_k a) \equiv r_1 r_2 \cdots r_k = s_1 s_2 \cdots s_k \pmod{b}.$$

Comparing, we see that

$$(s_1 s_2 \cdots s_k) a^k \equiv s_1 s_2 \cdots s_k \pmod{b},$$

or

$$b \mid (s_1 s_2 \cdots s_k)(a^k - 1).$$

Since $s_1 s_2 \ldots s_k$ is relatively prime to b, this implies that $b \mid a^k - 1$ as claimed.

6.10 How to Test Whether a Number is a Prime

6.10.1. By induction on k. True if $k = 1$. Let $n = 2m + a$, where a is 0 or 1. Then m has $k - 1$ bits, so by induction, we can compute 2^m using at most $2(k-1)$ multiplications. Now $2^n = (2^m)^2$ if $a = 0$ and $2^n = (2^m)^2 \cdot 2$ if $a = 1$.

6.10.2. If $3 \mid a$, then clearly $3 \mid a^{561} - a$. If $3 \nmid a$, then $3 \mid a^2 - 1$ by Fermat, hence $3 \mid (a^2)^{280} - 1 = a^{560} - 1$. Similarly, if $11 \nmid a$, then $11 \mid a^{10} - 1$ and hence $11 \mid (a^{10})^{56} - 1 = a^{560} - 1$. Finally, if $17 \nmid a$, then $17 \mid a^{16} - 1$ and hence $17 \mid (a^{16})^{35} - 1 = a^{560} - 1$.

7 Graphs

7.1 Even and Odd Degrees

7.1.1. There are 2 graphs on 2 nodes, 8 graphs on 3 nodes (but only four "essentially different"), 64 graphs on 4 nodes (but only 11 "essentially different").

7.1.2. (a) No; the number of odd degrees must be even. (b) No; node with degree 5 must be connected to all other nodes, so we cannot have a node with degree 0. (c) 12 (but they are all "essentially the same"). (d) $9 \cdot 7 \cdot 5 \cdot 3 \cdot 1 = 945$ (but again they are all "essentially the same").

7.1.3. This graph (which we will call a *complete graph*) has $\binom{n}{2}$ edges if it has n nodes.

7.1.4. In graph (a), the number of edges is 17, the degrees are $9, 5, 3, 3, 2, 3, 1, 3, 2, 3$. In graph (b), the number of edges is 31, the degrees are $9, 5, 7, 5, 8, 3, 9, 5, 7, 4$.

7.1.5. $\binom{10}{2} = 45$.

7.1.6. $2^{\binom{20}{2}} = 2^{190}$.

7.1.7. *Every graph has two nodes with the same degree.* Since each degree is between 0 and $n - 1$, if all degrees were different, then they would be $0, 1, 2, 3, \ldots n - 1$ (in some order). But the node with degree $n - 1$ must be connected to all the others, in particular to the node with degree 0, which is impossible.

7.2 Paths, Cycles, and Connectivity

7.2.1. There are 4 paths, 6 cycles, and 1 complete graph.

7.2.2. The edgeless graph on n nodes has 2^n subgraphs. The triangle has 18 subgraphs.

7.2.3. The path of length 3 and the cycle of length 5 are the only examples. (The complement of a longer path or cycle has too many edges.)

7.2.4. Yes, the proof remains valid.

7.2.5. (a) Delete any edge from a path. (b) Consider two nodes u and v. the original graph contains a path connecting them. If this does not go through e, then it remains a path after e is deleted. If it goes through e, then let $e = xy$, and assume that the path reaches x first (when traversed from u to v). Then after e is deleted, there is a path in the remaining graph from u to x, and also from x to y (the remainder of the cycle), so there is one from u to y. But there is also one from y to v, so there is also a path from u to v.

7.2.6. (a) Consider a shortest walk from u to v; if this goes through any nodes more than once, the part of it between two passes through this node can be deleted, to make it shorter. (b) The two paths together form a walk from a to c.

7.2.7. Let w be a common node of H_1 and H_2. If you want a path between nodes u and v in H, then you can take a path from u to w, followed by a path from w to v, to get a walk from u to w.

7.2.8. Both graphs are connected.

7.2.9. The union of this edge and one of these components would form a connected graph that is strictly larger than the component, contradicting the definition of a component.

7.2.10. If u and v are in the same connected component, then this component, and hence G too, contains a path connecting them. Conversely, if there is a path P in G connecting u and v, then this path is a connected subgraph, and a maximal connected subgraph containing P is a connected component containing u and v.

7.2.11. Assume that the graph is not connected and let a connected component H of it have k nodes. Then H has at most $\binom{k}{2}$ edges. The rest of the graph has at most $\binom{n-k}{2}$ edges. Then the number of edges is at most

$$\binom{k}{2} + \binom{n-k}{2} = \binom{n-1}{2} - (k-1)(n-k-1) \le \binom{n-1}{2}.$$

7.13 Eulerian Walks and Hamiltonian Cycles

7.3.1. The upper left graph does not have an Eulerian walk. The lower left graph has an open Eulerian walk. The two graphs on the right have closed Eulerian walks.

7.3.2. Every node with an odd degree must be the endpoint of one of the two walks, so a necessary condition is that the number of nodes with odd degree to be at most four. We show that this condition is also sufficient. We know that the number of nodes with odd degree is even. If this number is 0 or 2, then there is a single Eulerian walk (and we can take any single node as the other walk).

Suppose that this number is four. Add a new edge connecting two of the nodes with odd degree. Then there are only two nodes with odd degree left, so the graph has an Eulerian walk. Deleting the edge splits this walk into two, which together use every edge exactly once.

7.3.3. The first graph does; t~~he~~ second does not.

8 Trees

8.1 How to Define a Tree

8.1.1. If G is a tree, then it contains no cycles (by definition), but adding any new edge creates a cycle (with the path in the tree connecting the endpoints of the new edge). Conversely, if a graph has no cycles but adding any edge creates a cycle, then it is connected (either two nodes u and v are connected by an edge, or else adding an edge connecting them creates a cycle, which contains a path between u and v in the old graph), and therefore it is a tree.

8.1.2. If u and v are in the same connected component, then the new edge uv forms a cycle with the path connecting u and v in the old graph. If joining u and v by a new edge creates a cycle, then the rest of this cycle is a path between u and v, and hence u and v are in the same component.

8.1.3. Assume that G is a tree. Then there is at least one path between two nodes, by connectivity. But there cannot be two paths, since then we would get a cycle (find the node v where the two paths branch away, and follow the second path until it hits the first path again; follow the first path back to v to get a cycle).

Conversely, assume that there is a unique path between each pair of nodes. Then the graph is connected (since there is a path) and cannot contain a cycle (since two nodes on the cycle would have at least two paths connecting them).

8.2 How to Grow Trees

8.2.1. Start the path from a node of degree 1.

8.2.2. Any edge has only one lord, since if there were two, they would have to start from different ends, and they would have then two ways to get to the king: either continuing as they started, or waiting for the other and walking together. Similarly, an edge with no lord would have to lead to two different ways of walking.

8.2.3. Start at any node v. If one of the branches at this node contains more than half of all nodes, move along the edge leading to this branch. Repeat. You'll never backtrack, because this would mean that there is an edge whose deletion results in two connected components, both containing more than half of the nodes. You'll never cycle back to a node already seen, because the graph is a tree. Therefore, you must get stuck at a node such that each branch at this node contains at most half of all nodes.

8.3 How to Count Trees

8.3.1. The number of unlabeled trees on $2, 3, 4, 5$ nodes is $1, 1, 2, 3$. They give rise to a total of $1, 3, 16, 125$ labeled trees.

8.3.2. There are n stars and $n!/2$ paths on n nodes.

8.4 How to Store Trees

8.4.1. The first is the father code of a path; the third is the father code of a star. The other two are not father codes of trees.

8.4.2. This is the number of possible father codes.

8.4.3. Define a graph on $\{1, \ldots, n\}$ by connecting all pairs of nodes in the same column. If we do it backwards, starting with the last column, we get a procedure for growing a tree by adding a new node and an edge connecting it to an old node.

8.5.1. (a) encodes a path; (b) encodes a star; (c) does not encode any tree (there are more 0's than 1's among the first 5 elements, which is impossible in the planar code of any tree).

9 Finding the Optimum

9.1 Finding the Best Tree

9.1.1. Let H be an optimal tree and let G be the tree constructed by the pessimistic government. Look at the first step at which an edge $e = uv$ of H is eliminated. Deleting e from H we get two components; since G is connected, it has an edge f connecting these two components. The edge f cannot be more expensive than e, or else the pessimistic government would have chosen f to eliminate instead of e. But then we can replace e by f in H without increasing its cost. Hence we conclude as in the proof given above.

9.1.2. Take nodes $1, 2, 3, 4$ and costs $c(12) = c(23) = c(34) = c(41) = 3$, $c(13) = 4$, $c(24) = 1$. The pessimistic government builds (12341), while the best solution is 12431.

9.2 Traveling Salesman

9.2.1. No, because it intersects itself (see next exercise).

9.2.2. Replacing two intersecting edges by two other edges pairing up the same 4 nodes, just differently, gives a shorter tour by the triangle inequality.

10 Matchings in graphs

10.1 A Dancing Problem

10.1.1. If every degree is d, then the number of edges is $d \cdot |A|$, but also $d \cdot |B|$.

10.1.2. (a) a triangle; (b) a star.

10.1.3. A graph in which every node has degree 2 is the union of disjoint cycles. If the graph is bipartite, these cycles have even length.

10.3 The Main Theorem

10.3.1. Let $X \subseteq A$ and let Y denote the set of neighbors of X in B. There are exactly $d|X|$ edges starting from X. Every node in Y accommodates no more than d of these; hence $|Y| \geq |X|$.

10.3.2. The assumption for $X = A$ yields that $|B| \geq |A|$. If $|B| = |A|$, then we already know the assertion (Theorem 10.3.1), so suppose that $|B| > |A|$. Add $|B| - |A|$ new nodes to A to get a set A' with $|A'| = |B|$. Connect every new node to every node in B. The graph we get satisfies the conditions in the Marriage Theorem (Theorem 10.3.1): We have $|A'| = |B|$, and if $X \subseteq A'$ then either $X \subseteq A$ (in which case it has at least $|X|$ neighbors in B by the assumption of the exercise), or X contains a new node, in which case every node in B is a neighbor of X. So the new graph has a perfect matching. Deleting the newly added nodes, we see that the edges of the perfect matching that remain match all nodes of A with different nodes of B.

10.4 How to Find a Perfect Matching

10.4.1. On a path with 4 nodes, we may select the middle edge.

10.4.2. The edges in the greedy matching M must meet every edge in G (otherwise, we could further extend M), in particular every edge in the perfect matching matching. So every edge in the perfect matching has at most one endpoint unmatched by M.

10.4.3. The largest matching has 5 edges.

10.4.4. If the algorithm terminates without a perfect matching, then the set S shows that the graph is not "good."

11 Combinatorics in Geometry

11.1 Intersections of Diagonals

11.1.1. $\frac{n(n-3)}{2}$.

11.2 Counting Regions

11.2.1. True for $n = 1$. Let $n > 1$. Delete any line. The remaining lines divide the plane into $(n - 1)n/2 + 1$ regions by the induction hypothesis. The last line cuts n of these into two. So we get

$$\frac{(n - 1)n}{2} + 1 + n = \frac{n(n + 1)}{2} + 1.$$

11.3 Convex Polygons

11.3.1. See Figure 16.1.

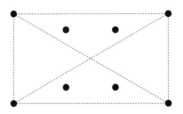

FIGURE 16.1.

12 Euler's Formula

12.1 A Planet Under Attack

12.1.1. There are n nodes of degree $n - 1$ and $\binom{n}{4}$ nodes of degree 4 (see Section 11.1). So the number of edges is $\frac{1}{2}\left(n \cdot (n-1) + \binom{n}{4} \cdot 4\right)$. From Euler's Formula, the number of countries is

$$\left(2\binom{n}{4} + \binom{n}{2}\right) - \left(n + \binom{n}{4}\right) + 2 = \binom{n}{4} + \binom{n}{2} - n + 2;$$

you have to subtract 1 for the country outside.

12.1.2. Let f be the number of regions of the island. Consider the graph formed by the dams and also the boundary of the island. There are $2n$ nodes of degree 3 (along the shore), and $\binom{n}{2}$ nodes of degree 4 (the intersection points of straight dams). So the number of edges is

$$\frac{1}{2}\left((2n) \cdot 3 + \binom{n}{2} \cdot 4\right) = 2\binom{n}{2} + 3n.$$

The number of countries is $f+1$ (we have to count the ocean too), so Euler's formula gives $f + 1 + 2n + \binom{n}{2} = 2\binom{n}{2} + 3n + 2$, whence $f = \binom{n}{2} + n + 1$.

12.2 Planar Graphs

12.2.1. Yes, see Figure 16.2.

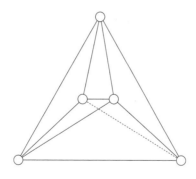

FIGURE 16.2.

12.2.2. No; the argument is similar to the one showing that K_5 is not planar. The houses, wells, and paths form a bipartite graph with 6 nodes and 9 edges. Suppose that this can be drawn in the plane without intersections. Then we have $9 + 2 - 6 = 5$ regions. Each region has at least 4 edges (since there are no triangles), and hence the number of edges is at least $\frac{1}{2} \cdot 5 \cdot 4 = 10$, which is a contradiction.

13 Coloring Maps and Graphs

13.1 Coloring Regions with Two Colors

13.1.1. By induction. True if $n = 1$. Let $n > 1$. Assume that the description of the coloring is valid for the first $n - 1$ circles. If we add the nth, the color and the parity don't change outside this circle; both change inside the circle. So the description remains valid.

13.1.2. (a) By induction. True for 1 line. Adding a line, we recolor all regions on one side.

(b) One possible description: Designate a direction as "up." Let p any point not on any of the lines. Start a half-line "up" from P. Count how many of the given lines intersect it. Color according to the parity of this intersection number.

13.2 Coloring Graphs with Two Colors

13.2.1. This graph cannot contain any odd cycle. Indeed, if we consider any cycle C, then each edge of it contains exactly one intersection point with the union of circles. The contribution of every circle is even, since walking around C, we cross the circle alternatingly in and out.

13.3 Coloring Graphs with Many Colors

13.3.1. Suppose that we have a good 3-coloring of the first graph. Starting from above, the first vertex gets (say) color 1, the vertices on the second level must get colors 2 and 3, and then both of the lowest two vertices must get color 1. But this is impossible, since they are connected.

Suppose that we have a good 3-coloring of the second graph. Starting from the center, we may assume that it has color 1, so its neighbors get colors 2 or 3. Now recolor each outermost vertex with color 1 by giving it the color of its inner "twin". This coloring would give a good coloring of a 5-cycle by 2 colors, since "twins" have the same neighbors (except that the inner twin is also connected to the center). This is a contradiction.

13.3.2. By rotating the plane a little, we may assume that all intersection points have different y coordinates (which we just call "heights"). Starting with the highest intersection point, and moving down, we can color the intersection points one by one. Each time, there are at most two intersection points that are adjacent to the current point along the two lines that were colored previously, and so we can find a color for the current point different from these.

13.3.3. We may assume that there are at least 2 nodes, and so there is a node of degree at most d. We delete it, recursively color the remaining graph by $d + 1$ colors, and then we can extend this coloring to the last point, since its d neighbors exclude only d colors.

13.3.4. We delete a point of degree d, and recursively color the remaining graph with $d + 1$ colors. We can extend this as in the previous solution.

14 Finite Geometries, Codes, Latin Squares, and Other Pretty Creatures

14.1 Small Exotic Worlds

14.1.1. The Fano plane itself.

14.1.2. Let abc be a circle. Then two of the lines through a contain b and c, respectively, so they are not tangents. The third line through a is the tangent.

14.1.3. If H is a hypercircle, then its 4 points determine 6 lines, and 3 of these 6 lines go through each of its points. So the seventh line does not go through any of the 4 points of the hypercircle. Conversely, if L is a line, then the 4 points not on L cannot contain another line (otherwise, these two lines would not intersect), and so these 4 points form a hypercircle.

14.1.4. (a) If everybody on line L votes yes, then (since every line intersects L) every line has at least one point voting yes, and so no line will vote all no. (b) We may assume that at least 4 points vote yes; let a, b, c, and d be 4 of them. Suppose that there is no line voting all yes. Then each of the 3 lines through a contains at most one further yes vote, so each of them must contain exactly one of b, c, and d. So the remaining 3 points vote no. The yes votes form a hypercircle (exercise 14.1.3), so the no votes form a line.

14.1.5. (a) Through two original points there is the original line; through an original point a and a new point b there is a unique line through a among all parallel lines to which b was added; and for two new points there is the new line. (b) is similar. (c) is obvious. (d) follows from (a), (b), and (c), as we saw above.

14.1.6. Yes, for every line (2 points) there is exactly one line that is disjoint from it (the other 2 points).

14.1.7. See Figure 16.3 (there are many other ways to map the points).

14.1.8. This is not a coincidence. Fix any point A of the Cube space. Every plane through A contains 3 lines through A. If we call the lines through a given point "POINTS," and those triples of these lines that belong to one plane "LINES," then these POINTS and LINES form a Fano plane.

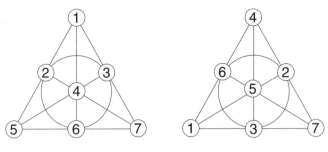

FIGURE 16.3.

14.2 Finite Affine and Projective Planes

14.2.1. Fix any point a. There are $n + 1$ lines through a, which have no other points in common and cover the whole plane by (a). Each of these lines has n points besides a, so there are $(n + 1)n$ points besides a, and $n(n + 1) + 1 = n^2 + n + 1$ points altogether.

14.2.2. We can assign coordinates to the vertices of the cube as if it were in Euclidean space, but think of the coordinates as elements of the 2-element field (Figure 16.4). Then it is straightforward (if lengthy) to check that the planes of the Cube space are precisely the sets of points given by linear equations. For example, the linear equation $x + y + z = 1$ gives the points 001, 010, 011, 111 (don't forget that we are working in the 2-element field), which is just the plane consisting of the light points.

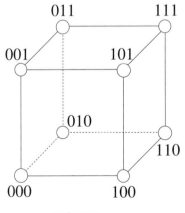

FIGURE 16.4.

14.2.3. A projective plane of order 10 ought to have $10^2 + 10 + 1 = 111$ points, 111 lines, with 11 points on each line. The number of ways to select a candidate line is $\binom{111}{11}$; the number of ways to select 111 candidate lines is

$$\binom{\binom{111}{11}}{111} = \binom{473239787751081}{11} > 10^{1448}.$$

One could not check so many possibilities even with the fastest computer within the lifetime of the universe! Lam, Thiel, and Swiercz had to work in a much more sophisticated manner.

14.3 Block Designs

14.3.1. 441, 44.

14.3.2. For any two citizens C and D, there are λ clubs containing both. If we add this up for every D, we count $(v-1)\lambda$ clubs containing C. Each such club is counted $k-1$ times (once for every member different from C, so the number of clubs containing C is $(v-1)\lambda/k$. This is the same for every citizen C.

14.3.3. (a) $v = 6$, $r = 3$, $k = 3$ gives $b = 6$ by (14.1), but $\lambda = 6/5$ by (14.2). (b) $b = 8$, $v = 16$, $r = 3$, $k = 6$, $\lambda = 1$ (there are many other examples in both cases).

14.3.4. Take $b = v$ clubs, and construct for every citizen C a club in which everybody else is a member except C. Then $b = v$, $k = v - 1$, $r = v - 1$, $\lambda = v - 2$.

14.4 Steiner Systems

14.4.1. Let A, B, C be 3 elements that do not form a club. There is a unique club containing A and B, which has a unique third element; call this D. Similarly, there is a unique element E such that ACE is a club, and a unique element F such that BCF is a club. The elements D, E, F must be distinct, since if (say) $D = E$, then A and D are contained in two clubs (one with B and one with C). Let the seventh element be G. There is a unique club containing C and D, and the third member of this club must be G (we can check that any of the other 4 choices would yield two clubs with two members is common). Similarly, AFG and BEG are clubs. Similarly, there is a unique club containing D and E, whose third member must be F. So, apart from the names of the citizens, the club structure is uniquely determined.

14.4.2. We have $r = (v-1)/2$ by (14.2), and hence $b = v(v-1)/6$ by (14.1). Since $v - 1 \geq 6$, we have $b \geq v$.

14.4.3. Call a triple contained in S an S-*triple*. The total number of triples is $b = v(v-1)/6$, the number of S-triples is

$$b' = \frac{\frac{v-1}{2}\left(\frac{v-1}{2} - 1\right)}{6} = \frac{(v-1)(v-3)}{24},$$

and so the number of non-S-triples is $b - b' = \frac{(v+1)(v-1)}{8}$. Every non-$S$-triple has at most one point in S and thus at least two points not in S. But the number of pairs not is S is $\binom{(v+1)/2}{2} = \frac{(v+1)(v-1)}{8}$, and since these pairs can belong to one of the non-S-triples only, it follows that each of the non-S-triples must contain exactly one pair of elements outside S. This proves that each non-S-triple must contain an element of S.

14.4.4. See Figure 16.5.

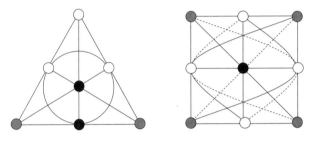

FIGURE 16.5.

14.4.5. Every girl has 8 other girls to walk with; every day she can walk with 2 in a line. So 4 days are necessary for her to walk with everybody exactly once.

14.5 Latin Squares

14.5.1. There are 576 different 4×4 Latin squares. There are many ways to arrive at this figure; we sketch one. The first row can be filled out 4! ways. These are all equivalent in the sense that the number of ways they can be completed is the same for each of them, so we may fix the first row as 0 1 2 3 and just count the number of ways to complete this. The first column now can be filled out in 3! ways, and again all of these are equivalent, so let's fix it as 0 1 2 3.

If the 0 in the second row is in the second position, then the rest of the second row and second column is forced, but we get two ways to fill out the 4 fields in the lower right corner. If the 0 in the second row is in the third or fourth position, then the way to fill out the rest is forced. Thus we get the 4 Latin squares below:

```
0 1 2 3      0 1 2 3      0 1 2 3      0 1 2 3
1 0 3 2      1 0 3 2      1 2 3 0      1 3 0 2
2 3 0 1      2 3 1 0      2 3 0 1      2 0 3 1
3 2 1 0      3 2 0 1      3 0 1 2      3 2 1 0
```

Therefore, the number of ways to fill out the remaining 9 fields is 4, so the total number is $4! \cdot 3! \cdot 4 = 576$.

These four may look different, but if we flip 1 and 2 in the third, flip rows 2 and 3, and flip columns 2 and 3, we get the second. Similarly, if we flip 1 and 3 in the fourth, flip rows 2 and 4, and flip columns 2 and 4, we get the second. So the last 3 of these are not essentially different.

There is no way to get the second square from the first by such operations (this follows, e.g., by Exercise 14.5.7). So there are two essentially different 4×4 Latin squares.

14.5.2. This is quite simple. For example, the table below is good (there are many other possibilities):

0	1	2	$\ldots$	$n-2$	$n-1$
1	2	3	$\ldots$	$n-1$	0
2	3	4	$\ldots$	0	1
$\vdots$					$\vdots$
$n-1$	0	1	$\ldots$	$n-3$	$n-2$

14.5.3.

1	2	3		1	2	3
2	3	1		3	1	2
3	1	2		2	3	1

14.5.4. We add 1 to every number; in this way, every row and column sum increases by 4.

14.5.5. We need two Latin squares where not only in the rows and columns, but also in the diagonals every number occurs once. These two will do:

0	1	2	3		0	1	2	3
2	3	0	1		3	2	1	0
3	2	1	0		1	0	3	2
1	0	3	2		2	3	0	1

From these two we get the following perfect magic square:

0	5	10	15
11	14	1	4
13	8	7	2
6	3	12	9

14.5.6. If there exists such a Latin square, then arbitrarily permuting the numbers $0, 1, 2, 3$ in it would give another square orthogonal to the three squares in (14.9) and (14.12). (Prove!) So we may start with a square having its first row 0 1 2 3. But then what can its first entry in the second row be? Zero is impossible (because the entry above it is also 0), but 1, 2, or 3 are also ruled out: for instance, if we had 2, then it wouldn't be orthogonal to square (14.12), because the pair $(2, 2)$ would occur twice. So there does not exist such a Latin square. (Try to generalize this result: From the $n \times n$

Latin squares we can choose at most $n - 1$ squares pairwise orthogonal to each other.)

14.5.7. If we had a square orthogonal to (14.13), then using the same argument as in the solution of Exercise 14.5.6, we may suppose that the first row is 0 1 2 3. Then the pairs $(0,0)$, $(1,1)$, $(2,2)$, and $(3,3)$ occur in the first row, which implies that in the other rows, the two squares cannot have the same number in the same position.

In particular, the first entry of the second row cannot be 1, and it cannot be 0 (because the entry above it is 0). So it is 2 or 3.

Suppose it is 2. Then the second entry in this row cannot be 1 or 2 (there is a 1 above it and a 2 before it), and it cannot be 3, so it is 0. The fourth entry cannot be 2, 0 or 3, so it must be 1; it follows that the second row is 2 0 3 1 (the same as the third row in (14.13)). Next we can figure out the last row: Each entry is different from the two above it in the first and second row, and also from the last row of (14.13), which implies that this row must be the same as the second row of (14.13): 1 3 0 2. Hence the third row must be 3 2 1 0; but now the pair $(3,1)$ occurs twice when the last two rows are overlaid.

The case where the second row starts with a 3 can be argued in the same way.

14.6 Codes

14.6.1. Suppose that a code is d-error-correcting. We claim that for any two codewords we must flip at least $2d+1$ bits to get from one to the other. Indeed, if we could get from codeword u to codeword v by flipping only $2d$ bits, then consider the codeword w obtained from u by flipping d of these bits. We could receive w instead of u, but also instead of v, so the code is not d-error-correcting.

Now if we receive any message that has at most $2d$ errors, then this message is not another codeword, so we can detect up to $2d$ errors.

The converse is proved similarly.

14.6.2. If the string has no 1's, then it is a codeword. If it contains one 1, this can be flipped to get a codeword. If it has two 1's, then there is a line through the corresponding two points of the Fano plane, and flipping the 0 in the position corresponding to the third point gives a codeword. If it has three 1's, and these are collinear, then it is a codeword. If it has three 1's, and these are not collinear, then there is a unique point not on any of the three lines determined by them, and flipping this we get a codeword. If it contains at least four 1's, then we can argue similarly, interchanging the role of 1's and 0's.

15 A Glimpse of Complexity and Cryptography

15.2 Classical cryptography

15.2.1. I THINK WE SHOULD NOT ATTACK FOR ANOTHER WEEK, BUT THEN WITH FULL FORCE. BELA

15.2.2. Let $a_1 a_2 \ldots a_n$ be the key and $b_1 b_2 \ldots b_n$ the plain text. Caligula intercepts one message whose bits are $a_2 \oplus b_1, a_3 \oplus b_2, \ldots a_n \oplus b_{n-1}$, and another message whose bits are $a_1 \oplus b_1, a_2 \oplus b_2, \ldots, a_n \oplus b_n$. (The second message is one bit longer, which may give him a hint of what happened.) He can compute the binary sum of the first bits, second bits, etc. So he gets $(a_2 \oplus b_1) \oplus (a_1 \oplus b_1) = a_1 \oplus a_2$, $(a_3 \oplus b_2) \oplus (a_2 \oplus b_2) = a_2 \oplus a_3$, etc.

Now he guesses that $a_1 = 0$; since he knows $a_1 \oplus a_2$, he can compute a_2, then similarly a_3, and so on, until he gets the whole key. It may be that his initial guess was wrong, which he notices, since in trying to decode the message he gets garbage; but then he can try out $a_1 = 1$, and recover the key. One of the two guesses will work.

15.2.3. Let $a_1 a_2 \ldots a_n$ be the key and let $b_1 b_2 \ldots b_n$ and $c_1 c_2 \ldots c_n$ be the two plain texts. Caligula intercepts one message whose bits are $a_1 \oplus b_1, a_2 \oplus b_2, \ldots a_n \oplus b_n$, and another message whose bits are $a_1 \oplus c_1, a_2 \oplus c_2, \ldots, a_n \oplus c_n$. As before, he computes the binary sum of the first bits, second bits, etc., to get $(a_1 \oplus b_1) \oplus (a_1 \oplus c_1) = b_1 \oplus c_1$, $(a_2 \oplus b_2) \oplus (a_2 \oplus c_2) = b_2 \oplus c_2$, etc.

The rest is not as straightforward as in the previous exercise, but suppose that Caligula can guess part of (say) Arthur's message (signature, or address, or the like). Then, since he knows the bit-by-bit binary sum of the two messages, he can recover the corresponding part of Bela's message. With luck, this is not a full phrase, and it contains part of a word. Then he can guess the rest of the word, and this gives him a few more letters of Arthur's message. With luck, this suggests some more letters of Bela's message, etc.

This is not completely straightforward, but typically it gives enough information to decode the messages (as World War II codebreakers learned). One important point: Caligula can *verify* that his reconstruction is correct, since in this case both messages must turn out to be meaningful.

15.3 How to Save the Last Move in Chess

15.3.1. Alice can easily cheat: She can send just a random string x in the evening, figure out her move overnight, along with the string y that encodes it, and send the binary sum of x and y as the alleged key.

15.3.2. This certainly eliminates the cheating in the previous exercise, since if she changes her mind, the "key" she computes back from the message the next morning will not be meaningful. But now Bob has the advantage: He can try out all "random but meaningful" keys, since there are not so many of them.

15.6 Public Key Cryptography

15.6.1. (a) Pick random numbers (public keys) $e_1, e_2, \ldots e_M$ and apply the hypothesized algorithm to compute the the corresponding secret keys $d_1, d_2, \ldots, d_M$. The number $k = (p-1)(q-1)$ is a common divisor of $e_1 d_1 - 1, e_2 d_2 - 1, \ldots, e_M d_M - 1$, so it is a divisor of $K = \gcd(e_1 d_1 - 1, e_2 d_2 - 1, \ldots, e_M d_M - 1)$, which we can compute. If $K < m$, then we know that in fact $k = K$, since $k = (p-1)(q-1) > pq/2 = m/2$. Otherwise we pick another public key e_{M+1} and repeat. One can show that after no more than about $\log m$ iterations, we find k with large probability.

(b) If we know $m = pq$ and $k = (p-1)(q-1)$, then we know $p+q = m-k+1$, and so p and q can be determined as the solutions of the quadratic equation $x^2 - (m - k + 1)x + m = 0$.

Index

Undergraduate Texts in Mathematics

(continued from page ii)

Gamelin: Complex Analysis.

Gordon: Discrete Probability.

Hairer/Wanner: Analysis by Its History. *Readings in Mathematics.*

Halmos: Finite-Dimensional Vector Spaces. Second edition.

Halmos: Naive Set Theory.

Hämmerlin/Hoffmann: Numerical Mathematics. *Readings in Mathematics.*

Harris/Hirst/Mossinghoff: Combinatorics and Graph Theory.

Hartshorne: Geometry: Euclid and Beyond.

Hijab: Introduction to Calculus and Classical Analysis.

Hilton/Holton/Pedersen: Mathematical Reflections: In a Room with Many Mirrors.

Hilton/Holton/Pedersen: Mathematical Vistas: From a Room with Many Windows.

Iooss/Joseph: Elementary Stability and Bifurcation Theory. Second edition.

Isaac: The Pleasures of Probability. *Readings in Mathematics.*

James: Topological and Uniform Spaces.

Jänich: Linear Algebra.

Jänich: Topology.

Jänich: Vector Analysis.

Kemeny/Snell: Finite Markov Chains.

Kinsey: Topology of Surfaces.

Klambauer: Aspects of Calculus.

Lang: A First Course in Calculus. Fifth edition.

Lang: Calculus of Several Variables. Third edition.

Lang: Introduction to Linear Algebra. Second edition.

Lang: Linear Algebra. Third edition.

Lang: Short Calculus: The Original Edition of "A First Course in Calculus."

Lang: Undergraduate Algebra. Second edition.

Lang: Undergraduate Analysis.

Laubenbacher/Pengelley: Mathematical Expeditions.

Lax/Burstein/Lax: Calculus with Applications and Computing. Volume 1.

LeCuyer: College Mathematics with APL.

Lidl/Pilz: Applied Abstract Algebra. Second edition.

Logan: Applied Partial Differential Equations.

Lovász/Pelikán/Vesztergombi: Discrete Mathematics.

Macki-Strauss: Introduction to Optimal Control Theory.

Malitz: Introduction to Mathematical Logic.

Marsden/Weinstein: Calculus I, II, III. Second edition.

Martin: Counting: The Art of Enumerative Combinatorics.

Martin: The Foundations of Geometry and the Non-Euclidean Plane.

Martin: Geometric Constructions.

Martin: Transformation Geometry: An Introduction to Symmetry.

Millman/Parker: Geometry: A Metric Approach with Models. Second edition.

Moschovakis: Notes on Set Theory.

Owen: A First Course in the Mathematical Foundations of Thermodynamics.

Palka: An Introduction to Complex Function Theory.

Pedrick: A First Course in Analysis.

Peressini/Sullivan/Uhl: The Mathematics of Nonlinear Programming.

Prenowitz/Jantosciak: Join Geometries.

Priestley: Calculus: A Liberal Art. Second edition.

Protter/Morrey: A First Course in Real Analysis. Second edition.

Protter/Morrey: Intermediate Calculus. Second edition.

Pugh: Real Mathematical Analysis.

Undergraduate Texts in Mathematics